Mathematics

for the IB Middle Years Programme

MYP Year 1

SERIES EDITOR: IBRAHIM WAZIR

MATTEO MERLO, DIANE OLIVER, KATHERINE PATE, MICHELLE SHAW

Published by Pearson Education Limited, 80 Strand, London, WC2R 0RL.

www.pearson.com/international-schools

Text © Pearson Education Limited 2021
Development edited by Eric Pradel
Edited by Sam Hartburn
Indexed by Georgina Bowden
Designed by Pearson Education Limited
Typeset by Tech-Set LTD
Picture research by SPi
Original illustrations © Pearson Education Limited 2021
Cover design © Pearson Education Limited 2021

The right of Matteo Merlo, Diane Oliver, Katherine Pate, Michelle Shaw and
Ibrahim Wazir to be identified as the authors of this work has been asserted by
them in accordance with the Copyright, Designs and Patents Act 1988.

First published 2021

24
10 9 8

British Library Cataloguing in Publication Data
A catalogue record for this book is available from the British Library

ISBN 978 1 292 36740 8

Printed and bound by CPI Group (UK) Ltd, Croydon CR0 4YY

Acknowledgements

(Key: b-bottom; c-centre; l-left; r-right; t-top)

Text Credits:

International Labour Organization: Global estimates of child labour: Results
and trends, 2012-2016, Geneva 2017, © International Labour Organization
324; **Statista:** Global water withdrawals per capita as of 2018, by select country,
© Statista 2020 335.

Prominent Photo Credits:

Shutterstock: Pikepicture/Shutterstock CVR.

Non-Prominent Photo Credits:

123RF: Nataliya Yakovleva/Shutterstock 53; Taui/123RF 134; Lauris
Sirmais/123RF 181; Mark Herreid/123RF 215; Ammit/123RF 235;
Homestudio/123RF 282.

Getty Images: AbeSnap23/iStock/Getty Images 174; Karanik Yimpat/EyeEm/
Getty Images 321; C Squared Studios/Photodisc/Getty Images 227b; C Squared
Studios/Stockbyte/Getty Images 227t.

Shutterstock: Daimond Shutter/Shutterstock 1; Parinyabinsuk/Shutterstock
3; Ekkapon/Shutterstock 10; Jan Martin Will/Shutterstock 18; Gopixa/
Shutterstock 31; Rawpixel.com/Shutterstock 33; LADO/Shutterstock 36; Volha
Hlinskaya/Shutterstock 47; bluecrayola/Shutterstock 52; HeinzTeh/Shutterstock
55; Chalermpon Poungpeth/Shutterstock 69; Boxthedog/Shutterstock 75;
Maren Winter/Shutterstock 83; Cheryl Savan/Shutterstock 88; Mountainpix/
Shutterstock 104; Varavin88/Shutterstock 106; Andrew Zarivny/Shutterstock
117; Xavier gallego morell/Shutterstock 126; Zvone/Shutterstock 146; humdan/
Shutterstock 151; Nevada31/Shutterstock 162; Arsenii Palivoda/Shutterstock 169;
Andrey_Popov/Shutterstock 171; Images By Kenny/Shutterstock 172; Creative
Stock55/Shutterstock 176; Ethan Daniels/Shutterstock 187; NadyaEugene/
Shutterstock 189; Eugene Lu/Shutterstock 191; Forance/Shutterstock 194; Goran
Jakus/Shutterstock 196; Cegli/Shutterstock 218; Joseph Calev/Shutterstock 220;
TRMK/Shutterstock 249; Kckate16/Shutterstock 255; Lyudmila Zavyalova/
Shutterstock 256; Fizkes/Shutterstock 259; Praisaeng/Shutterstock 285;
BeautifulBlossoms/Shutterstock 288; ImageFlow/Shutterstock 297; Kyle Tunis/
Shutterstock 301; Pablo Eder/Shutterstock 314; JohnKwan/Shutterstock 315b;
Prachaya Roekdeethaweesab/Shutterstock 318l; Suwit Rattiwan/Shutterstock 320;
TeamDAF/Shutterstock 326.

NASA: What's the secret code for talking to spacecraft?, NASA 52.

WorldData.info: Average income around the world, © WorldData.info 312.

American Association for the Advancement of Science: New global study
reveals the 'staggering' loss of forests caused by industrial agriculture By Erik
Stokstad, September 13, 2018, © American Association for the Advancement of
Science 315t.

United States Environmental Protection Agency: United States Environmental
Protection Agency, National Overview: Facts and Figures on Materials, Wastes
and Recycling 318c.

International Labour Organization: Global Wage Report 2016/17,
© International Labour Organization 2016 324.

All other images © Pearson Education

Contents

Course Structure

Mathematics can be fun!

'I can't do it!'

Have you ever thought or exclaimed these words when stumped by a mathematics problem? I bet every one of us has said these words to themselves at least once in their lifetime. And not just because of a mathematics problem! In order to be engaged as a learner, regardless of age, we like to experience things in a fun and interactive way. Not only that, learning only happens when we leave our comfort zone.

This series is dedicated to the idea that mathematics can be (and is) fun.

Our mission is to accompany you, dear learner, out of your comfort zone and towards the joy of mathematics. Do you accept this challenge?

Bearing in mind the latest research about learning mathematics, the driving ideas behind this series are the following:

- We believe that everyone can do mathematics. Of course there are a few that find it 'easier' than others, *but mathematics learning*, done right, *is for all*. We believe in you, the learner.

- *The essence of mathematics is solving problems*. We will work together to help you become a better problem solver. Problem solvers make mistakes. Plenty of them. With perseverance, they end up solving their problems. You can too. Making mistakes and learning from them is part of our education. Our approach is backed by research on *growth mindsets* and follows in the steps of George Pólya, the father of problem solving.

- Other than some special inventions, most of our societies' development is done by groups. That is why we will, with the help of your teachers, support you to achieve your goals within a group environment.

- Mathematicians' work, no matter how 'advanced' the result, starts with an exploration. Ideas do not magically materialise to a mathematician's mind by superpowers. Mathematicians work hard, and while working, discoveries are made. Whoever discovered gravitational forces did not sit back and then all of a sudden come out with the idea. It was observation first.

- Once you have an idea, you can investigate it to develop your understanding in more depth. We have included many opportunities for you to expand your knowledge further.

How to use this book

No one can teach you unless you want to learn. We believe that, through this partnership with you, we can achieve our goals.

In this book, we have introduced each concept with an Explore. First and foremost, when you start a new concept, try to do the Explore. Have courage to make guesses but try to justify your guesses. Remember, it is ok to make mistakes. Work with others to analyse your mistakes.

Explore 9.4

Look back at the data for rolling a dice 36 times from Explore 9.3.

How would you represent this data to make it easy to read?

Throughout this book, you will find worked examples. When you are given a worked example, do not jump immediately to the solution offered. Try it yourself first. When you do look at the solution offered, be critical and ask yourself: could I have done it differently?

Fact

The legend is that Isaac Newton discovered gravity when he saw a falling apple while thinking about the forces of nature. Whatever really happened, Newton realised that some force must be acting on falling objects like apples because otherwise they would not start moving from rest.

It is also claimed that Indian mathematician and astronomer Brahmagupta-II (598–670) discovered the law of gravity over 1000 years before Newton (1642–1727) did. Others claim that Galileo discovered it 100 years before Newton.

Worked example 8.3

Marta wants to cut 1 metre of ribbon into three equal lengths.

How long should each length be?

Give your answer to a suitable degree of accuracy.

Solution

$100 \div 3 = 33.333\ldots$ cm

It is not possible to measure 33.333… cm accurately.

Each length is 33.3 cm (to the nearest millimetre).

At the end of any activity, we encourage you to reflect on what you have done. There is always a chance to extend what you have learned to new ideas or different perspectives. Not only in studying mathematics, but in any task you should always take the opportunity to reflect on what you did. You will either feel that the task is completed, or you may find that you need to improve on some parts of it. This is true whether you are a student, a teacher, a parent, an engineer, or a business leader, to mention a few. You will find reflection boxes throughout the book to help you with this.

 Reflect

In Worked example 8.2, how did it help to convert the measurements to cm?

Can you round 5.26 m to the nearest 10 cm without converting to cm first?

At the end of each section of the book, you will find practice questions. It is recommended that you do these, and more, until you feel confident that you have mastered the concept at hand.

 Practice questions 8.2

1   Write 7.517 metres as ____ m ____ cm ____ mm

2   Write 3 m 24 cm 5 mm:

    a   in centimetres, to the nearest mm

    b   in metres, to the nearest mm.

Instead of summarising each chapter for you, we have you review what you learned from the chapter in a self-assessment. These self-assessments are checklists. Look at them, and if you feel you missed something, revisit the section covering it.

 Self assessment

I can identify natural numbers, integers and real numbers.

I can identify and use the place value of digits in natural numbers up to hundreds of millions.

I can identify and use the place value of digits in decimals.

Finally, at the end of every chapter, it is good practice to look back at the chapter as a whole and see whether you can solve problems. Each chapter contains check your knowledge questions for this purpose.

? Check your knowledge questions

1 Complete these measurement conversions.

a 420 m = ☐ cm

b 530 cm = ☐ m

c 0.4 km = ☐ mm

d 546 mm = ☐ cm

e 2450 m = ☐ km

f 3.4 m = ☐ mm

2 How many books of width 18 mm will fit on a shelf 1 m long?

During your course, your teacher will help you work in groups. In group work, ask for help and help others when asked. The best way of understanding an idea is when you explain it to someone else.

Remember, mathematics is not a bunch of calculations. Mathematical concepts must be communicated clearly to others. Whenever you are performing a task, justify your work and communicate it clearly.

Additional features

Matched to the latest MYP Mathematics Subject Guide

Key concepts, related concepts and global contexts

Each chapter covers one key concept and one or more related concepts in addition to being set within a global context to help you understand how mathematics is applied in our daily lives.

KEY CONCEPT

Relationships

RELATED CONCEPTS

Patterns, Quantity, Representation, Systems

GLOBAL CONTEXT

Globalisation and sustainability

Statement of inquiry and inquiry questions

Each chapter has a statement of inquiry and inquiry questions that lead to the exploration of concepts. The inquiry questions are categorised as factual, conceptual and debatable.

Statement of inquiry

Using number systems allows us to understand relationships that describe our climate, so we are able to acknowledge human impact on global climate change.

Approaches to learning tags

We have identified activities and questions that have a strong link to specific approaches to learning to help you understand where you are using particular skills.

 Thinking skills

Do you recall?

At the start of each chapter, you will find do you recall questions to remind you of the relevant prior learning before you start a new chapter. Answers to the do you recall questions can be found in the answers section at the back of the book.

Do you recall?

1 What are directed numbers?

2 What are the number operations?

3 What mental methods do you know for adding two 2-digit numbers?

4 What mental methods do you know for subtracting from a 2-digit number?

Investigations

Throughout the book, you will find investigation boxes. These investigations will encourage you to seek knowledge and develop your skills. They will often provide an opportunity for you to work with others.

 Investigation 1.1

Collect magazine, newspaper or online articles that use global temperatures, sea levels and carbon dioxide emissions. Explain in each case what the data tells you.

Research temperatures and the amount of rainfall for five different locations on the same day or month each year for 20 years. What does your data show you?

Fact boxes

Fact boxes introduce historical or background information for interest and context.

 Fact

A pescatarian is someone who eats fish, but does not eat any other meat.

Hint boxes

Hint boxes provide tips and suggestions for how to answer a question.

 Hint Q9

Note the different units.

Reminders

These boxes are used to recap previous concepts or ideas in case you need a refresher.

Reminder

Always state how you have rounded the measurement in the final answer.

Connections

These boxes highlight connections to other areas of mathematics, or even other subjects.

 Connections

You learned how to measure and draw angles accurately in Chapter 4.

🏆 Challenge Q12

Challenge tags

We have identified challenging questions that will help you stretch your understanding.

This series has been written with inquiry and exploration at its heart. We aim to inspire your imagination and see the power of mathematics through your eyes.

We wish you courage and determination in your quest to solve problems along your your MYP mathematics journey. Challenge accepted.

Ibrahim Wazir, Series Editor

A note for teachers

Alongside the textbook series, we have also created digital Teacher Guides. These Guides include, amongst other things, ideas for group work, suggestions for organising class discussion using the Explores and detailed, customisable unit plans.

Number review
1

1 Number review

KEY CONCEPT

Relationships

RELATED CONCEPTS

Patterns, Quantity, Representation, Systems

GLOBAL CONTEXT

Globalisation and sustainability

Statement of inquiry

Using number systems allows us to understand relationships that describe our climate, so we are able to acknowledge human impact on global climate change.

Factual

- What are negative numbers?
- What is the order of operations?

Conceptual

- How do you add and subtract integers?

Debatable

- Why do we have directed numbers?
- Why is it important to have order of operations?

Do you recall?

1 What are directed numbers?
2 What are the number operations?
3 What mental methods do you know for adding two 2-digit numbers?
4 What mental methods do you know for subtracting from a 2-digit number?

1.1 Natural numbers, integers and real numbers

In this section, we will consider three sets of numbers: natural numbers, integers and real numbers.

The set of natural numbers is represented by the symbol $\mathbb{N}$. Natural numbers represent the number of elements in a set, so $\mathbb{N} = \{0, 1, 2, 3, \ldots\}$. Counting numbers are the first set of numbers you used as a child; they are $\{1, 2, 3, \ldots\}$.

The set of integers is represented by the symbol $\mathbb{Z}$. The set of all positive and negative whole numbers, and zero make up the integers.

The set of real numbers is represented by the symbol $\mathbb{R}$. Natural numbers and integers are subsets of the real numbers.

Children start with the counting numbers when they learn to count.

Explore 1.1

Here is a set of numbers:

$$\{-2, -1.5, -1, -0.5, 0, \frac{1}{4}, 0.5, \frac{9}{10}, 1, 1.5, 2\}$$

Can you distinguish which of these numbers are natural numbers, integers and real numbers?

Which numbers are in all three number sets?

What immediately tells you that a number is an integer, but not natural?

What immediately tells you that a number is real, but not an integer?

Fact

$\{0, 1, 2, 3, \ldots\}$ is also called the set of whole numbers.

Hint

The set of natural numbers is:
$\mathbb{N} = \{0, 1, 2, 3, 4, 5, \ldots\}$

The set of integers is:
$\mathbb{Z} = \{\ldots, -2, -1, 0, 1, 2, \ldots\}$

The set of real numbers, $\mathbb{R}$, consists of all numbers including negative, positive, fractions and decimals.

Worked example 1.1

Which number sets do the following numbers belong to?

$-2 \qquad -0.4 \qquad 0 \qquad \frac{3}{4} \qquad 32$

Solution

-2 is in the set of integers and the set of real numbers.

-0.4 is in the set of real numbers.

0 is in all three sets, $\mathbb{N}$, $\mathbb{Z}$ and $\mathbb{R}$.

$\frac{3}{4}$ is in the set of real numbers.

32 is in all three sets, $\mathbb{N}$, $\mathbb{Z}$ and $\mathbb{R}$.

Reflect

Look back at the previous example.
How can you tell which numbers are in all three sets?
How can you tell which numbers are in one set only?
How can you tell which numbers are in two sets?

Practice questions 1.1

1 State the next natural number:

   a   greater than 6002

   b   greater than 8999

   c   less than 456

   d   less than 12 450.

2 State the next integer:

   a   greater than 9231

   b   greater than −5487

   c   greater than 10 759

   d   less than −758

   e   less than 0

   f   less than 13 490.

3 What natural number is 10 less than 104?

4 What integer is 50 less than 15?

5 Look at this list of numbers:

$-7, 0.4, 53, \dfrac{4}{5}, 34.5, 21, -2, -4.75, 14\,451, \dfrac{423}{1000}$

From the list write down the numbers in the set of:

   a   natural numbers

   b   integers

   c   real numbers.

6 Write down whether each statement is true or false.

   a   0.5 is an integer.

   b $\dfrac{7}{10}$ is a real number.

   c   −2 is an integer.

   d   −100 is a natural number.

   e   12 345 is a real number.

   f   0.459 is a real number.

   g   −199 is an integer.

   h   −4.6 is an integer.

7 What is the smallest even, non-zero, natural number?

8   Write down the next two numbers in each pattern. Say whether the
    pattern belongs to ℕ, ℤ or ℝ.

    a   10, 11, 12, 13, …   b   −21, −19, −17, −15, …

    c   0.1, 0.3, 0.5, 0.7, …   d $\frac{3}{5}, \frac{6}{5}, \frac{9}{5}, \frac{12}{5}, …$

    e   −10, −5, 0, 5, …   f   12 400, 12 450, 12 500, 12 550, …

9   Write down a real number between:

    a   1 and 2   b $\frac{2}{5}$ and $\frac{3}{5}$

    c   21.7 and 21.8   d   −49 and −48

    e   0.1 and 0.15   f   −5.5 and −5.6.

10 Write down your own set of five:

    a   natural numbers   b   integers

    c   real numbers.

11 a   Write down the set of the first 20 natural numbers.

    b   Which numbers in the set can be written as the sum of two
        consecutive numbers?

12 What is the sum of the first 100 non-zero, natural numbers? Explain
    the quickest way to work this out.

13 Create a set of six integers where dividing any number in the set by any
    smaller number in the set always gives an integer answer. Explain how
    you create this set of integers.

Hint Q11b

Consecutive numbers are
numbers that follow on
from each other in order,
for example, 10 and 11.

Hint Q12

There is a quick way to
work this out without
adding all 100 numbers
separately.

🏆 Challenge Q11b

🏆 Challenge Q12

🏆 Challenge Q13

1.2   Place value

1.2.1 Place value in natural numbers

Explore 1.2

Write down a 5-digit number. Read your number out. Swap your number
with others and read out each other's numbers.

Do the same with 6- and 7-digit numbers.

What is the largest number you can write and read?

Using all of the digits 2, 5, 7, 1, 9 and 4, write down the smallest 6-digit number you can and read it out. What value does the digit 5 have in your number?

Write down the largest 6-digit number you can. Use the digit 7 at least once. What value does the digit 7 have in your number?

Place value is the value of a digit in a number. Here is a recap of some place value headings.

Place value headings	Thousands	Hundreds	Tens	Ones
Dienes blocks				
Each column is the column to its right multiplied by 10.	1000 = 100 × 10	100 = 10 × 10	10 = 1 × 10	1

Worked example 1.2

a Write down the number three million, six hundred and four thousand and twenty as a numeral.

b Write down the value of the 2 and the 6 in 76 820.

Solution

a Arranging the information into a table shows us where there are zeros.

Millions	Hundred-thousands	Ten-thousands	Thousands	Hundreds	Tens	Ones
3	6	0	4	0	2	0

The number is written as 3 604 020.

b The number has 5 digits, so the first digit is a ten-thousands digit. Arranging the information into a table will show us the place value of each digit.

Ten-thousands	Thousands	Hundreds	Tens	Ones
7	6	8	2	0

The value of the 2 is 20.

The value of the 6 is 6000.

Write down the largest number that you can say. Share your number with others. Who has the largest number? How did you compare your numbers?

 Practice questions 1.2.1

1   Write the number represented by each diagram.

a

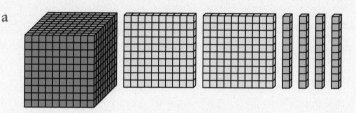

b

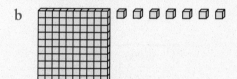

c

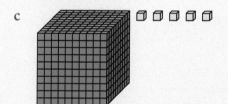

d

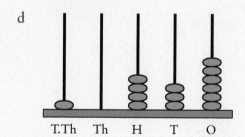

e

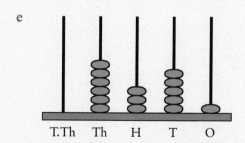

f

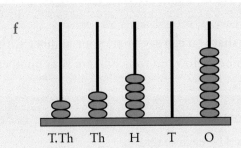

T.Th   Th   H   T   O

g
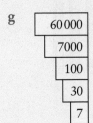

h
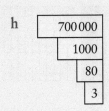

i
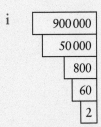

j
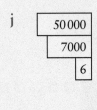

k
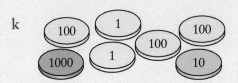

l
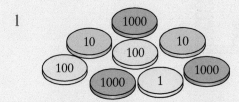

m
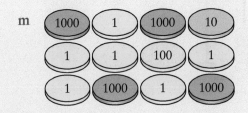

2   Write down each of the numbers given in words as numerals.

   a   two thousand, seven hundred and thirty-five

   b   four hundred and twelve thousand, nine hundred and seventeen

   c   three hundred and two thousand, one hundred and nine

   d   one million and twenty-four

   e   five million, eight hundred and forty-one thousand, two hundred
       and eighty-nine

   f   fifty-six million, six thousand, seven hundred and seventy

   g   nine hundred and twenty-three million, four hundred and forty-one
       thousand, seven hundred and eleven

   h   three hundred and eight million, five thousand and forty-nine.

3   Write down each numeral in words.

   a   7924   b   41 054   c   57 202   d   547 126

   e   908 040   f   8 545 782   g   59 025 118   h   20 105 067

4   Write down the value of the 3, the 4 and the 5 in each of these numerals.

   a   543   b   3045   c   32 405   d   53 024

   e   406 325   f   3 452 589   g   40 300 752   h   359 224 179

5   For each of these numbers, write down the value of each digit.

   a   658   b   9254   c   63 125   d   620 400

   e   900 218   f   5 470 816   g   59 025 618   h   670 809 243

6   Write each of these expanded numbers as numerals.

   a   20 000 + 8000 + 300 + 40 + 8

   b   700 000 + 30 000 + 2000 + 800 + 10 + 6

   c   5 000 000 + 800 000 + 10 000 + 2000 + 900 + 70 + 3

   d   9 000 000 + 60 000 + 7000 + 40 + 9

   e   9 000 000 + 600 000 + 7000 + 400 + 90

   f   5 000 000 + 80 000 + 1000 + 200 + 90 + 7

   g   60 000 000 + 1 000 000 + 80 000 + 900 + 30 + 7

   h   9 000 000 + 6000 + 500 + 3

Hint Q7

Expanded form means that the place value of each digit is written out in full. For example, 653 in expanded form is 600 + 50 + 3

Hint Q8

Ascending order means from the lowest value to the greatest value.

7   Write each of these numbers in expanded form.

a   49 523   b   54 870   c   870 252   d   7 521 463

8   Write each group of numbers in ascending order.

a   85 152, 83 658, 85 252, 84 125, 84 874

b   45 688, 45 025, 45 842, 45 000, 45 004

c   752 054, 749 254, 749 564, 749 887, 749 478

d   1 235 589, 1 259 471, 1 235 854, 1 238 008, 1 235 465

e   1 775 258, 1 897 429, 1 897 026, 1 036 987, 1 897 278

f   89 056, 87 981, 88 123, 88 598, 88 742

g   101 225, 100 000, 100 054, 101 568, 101 658, 100 471

h   6 127 554, 6 600 000, 6 541 965, 6 126 700

1.2.2 Place value in decimals

Explore 1.3

Using a decimal point and each of the digits 1, 2, 3, 4, 5, 6, 7, 8, 9 and 0 once only, what is the closest number you can make to 555.55?

Using the same digits, what is the closest number you can make to 800.9?

Using the same digits, what is the closest number you can make to 26 222.45?

Place value is the value of a digit in a number, including decimal numbers. Here is a recap of some place value headings.

The Chinese abacus was used for quick calculations involving large numbers hundreds of years before computers were invented. It relies on the concept of place value.

Place value headings	Ones		tenths	hundredths	thousandths
Each column is the column to its right multiplied by 10.	1	.	$\frac{1}{10} = 0.1$	$\frac{1}{100} = 0.01$	$\frac{1}{1000} = 0.001$

Worked example 1.3

a   Write 2.348 in expanded form.

b   Write 0.5, 0.05 and 0.055 in ascending order.

Solution

a Since we have 3 decimal places, the number has a thousandths digit. Arranging the information into a table will show us the place value of each digit.

Ones		tenths	hundredths	thousandths
2	.	3	4	8

$2.348 = 2 + 0.3 + 0.04 + 0.008$

b Ascending order means that the numbers are arranged from smallest to largest. We need to work out which is smallest, which is in the middle and which is largest.

We have three different numbers to compare. Each number has a different number of decimal places. Arranging the information into a table will show us the place value of each digit.

Ones		tenths	hundredths	thousandths
0	.	5	0	0
0	.	0	5	0
0	.	0	5	5

The table shows that 0.5 is the greatest value and 0.05 is the least value.

The numbers in ascending order are 0.05, 0.055, 0.5.

 Hint Qb

Use zeros to fill in blanks in the table. This will give each number the same number of decimal places.

 Fact

The decimal numeral system is an extension of the Hindu–Arabic numeral system.

 Reflect

Write down the smallest number that you can think of. Share your number with others. Who has the smallest number? How did you compare your numbers?

 Practice questions 1.2.2

1 Write down the value of the 7 in each of these numbers.

   a 0.79   **b** 0.972   **c** 0.0871

   d 7.154   **e** 58.573   **f** 6.0278

2 Write down the value of the 3 and the 4 in each of these numbers.

   a 3.4   **b** 3.04   **c** 5.34

   d 0.534   **e** 9.403   **f** 45.943

3 Write each of these expanded decimals as a single decimal.

    a   0.4 + 0.07 + 0.002   b   3 + 0.9 + 0.02 + 0.006

    c   0.5 + 0.09 + 0.004 + 0.0001   d   5 + 0.009 + 0.0005

    e   0.7 + 0.001 + 0.0006   f   9 + 0.07 + 0.00002

    g   40 + 5 + 0.9 + 0.008   h   60 + 7 + 0.04

4 Write each of these decimals in expanded form.

    a   0.785   b   2.56   c   0.7253   d   5.087

    e   0.6041   f   8.5026   g   24.8013   h   13.0205

5 Write down whether each statement is true or false.

    a   0.6 is the same value as 0.60

    b   0.3 is the same value as 0.03

    c   8 = 8.0

    d   0.07 = 0.7

    e   0.1 is the same value as 1.0

    f   0.002 = 0.0020

    g   0.03 = 0.30

6 Arrange these decimals in ascending order.

    a   0.4, 0.7, 0.3, 0.8

    b   3.11, 9.04, 1.9, 3.24

    c   0.7, 1, 0.33, 0.3, 0.71

    d   6, 0.6, 0.06, 6.6, 6.06

    e   0.4, 0.44, 0.04, 4.04, 4.4

    f   0.49, 0.51, 0.5, 0.4, 0.45

    g   0.75, 0.8, 0.684, 0.758, 0.81

    h   0.09, 0.1, 0.87, 0.8, 0.101

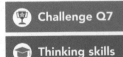

Challenge Q7

Thinking skills

7 Place the digits 0, 1, 2, 3, 4 in boxes ☐☐.☐☐☐ so that:

    a   3 has the greatest value in the number

    b   3 has the least value in the number

    c   your number is greater than 43.102

    d   your number is less than 40.132.

1.3 Number operations

1.3.1 Addition and subtraction

Addition and subtraction are inverse operations of each other. For example, adding 7 to 10 gives 17. Subtracting 7 brings you back to 10 again. The act of subtracting undoes the act of adding; this is what is meant by inverse operations.

 Explore 1.4

Can you work out the answer to each of these three calculations?

57 + 89   −5 + 9   23 − 50

For each calculation, explain how you worked out your answer.

Did you use the same method for each calculation?

Worked example 1.4

In a number pyramid, numbers in the lower layers determine the numbers above them. In this pyramid, each block in the upper layers can be found by adding together the two numbers below it.

Copy and complete the number pyramid.

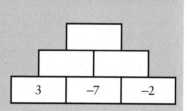

Solution

Adding 3 and −7 will give the value of the block directly above them. We can use the same method for the other blocks.

Work out 3 + (−7) and −7 + (−2) for the blocks on the middle row and then add the two answers together for the top block.

3 + (−7) = −4

−7 + (−2) = −9

−4 + (−9) = −13

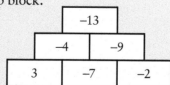

Hint

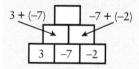

 Reflect

Look back at the number pyramid in Worked example 1.4.

If you are given the numbers on the bottom layer, how can you work out the top number without working out the middle layer?

If you change the order of the numbers on the bottom layer, will the top number change?

If you are given a target for the top number, how can you quickly find three possible numbers for the bottom?

Make a list of all the different words you can find that mean addition.
Make another list for all the words you can find that mean subtraction.

Connections

Game: Stick it!

This is a game for two players. Each player needs a copy of the empty sum.

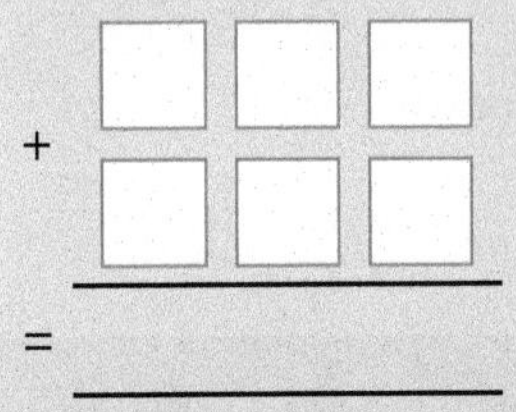

The target number is 695.

Each person rolls a dice and gets to place that number in one of their own blank spaces or in one of their opponent's blank spaces.

Your choice will either place yourself in a better position or place your opponent in a worse position. Play a few times and discuss your strategies with each other. You could change the target number.

Repeat this using 4-digit or 5-digit numbers. Do the same strategies apply?

Practice questions 1.3.1

1 Work out:

a   27 + 44

b   428 + 87

c   98 − 27

d   92 − 35

e   245 − 52

f   343 − 79

g   1248 + 2986

h   1358 − 89

i   234 + 45 + 27

j   1487 + 120 + 11

2   Here is a number pyramid. In a number pyramid, the value of any
    block is the sum of the two blocks directly below.
    Copy and complete this number pyramid.

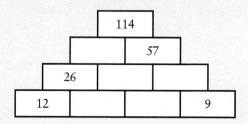

3   Work out:

    a   −3 + 5   b   −11 + 2

    c   −5 + 5   d   −12 + 8

    e   −3 + 15   f   −27 + 19

    g   −33 + 44   h   −100 + 35

4   Work out:

    a   2 − 5   b   10 − 15

    c   12 − 18   d   −1 − 9

    e   −5 − 5   f   −12 − 7

    g   −4 − 15   h   −27 − 12

    i   −33 − 44   j   −100 − 100

5   Work out:

    a   2 − (−5)   b   10 − (−8)

    c   2 − (−18)   d   1 − (−9)

    e   −8 − (−8)   f   −10 − (−7)

    g   −4 − (−15)   h   −17 − (−21)

    i   −44 − (−55)   j   −400 − (−100)

🛡 Hint Q5

Brackets can help make
sure that you do not
accidentally overlook one
of the minus signs, or
blur them into one when
writing hastily. Brackets
can also help to remind
you which buttons to enter
on your calculator.

2 − (−5) should be entered
as:

6 Work out:

a 62 + 25

b 102 − 48

c 13 − (−18)

d 10 − 29

e −15 − (−8)

f −48 − 7

g −4 − (−23)

h 177 − 29

i −100 − (−55)

j 427 − (−90)

 Challenge Q7

7 Here is a number pyramid. In a number pyramid, the value of any block is the sum of the two blocks directly below.
Copy and complete this number pyramid.

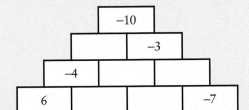

 Challenge Q8

8 In a magic square, all rows, columns and diagonals add up to the same number, known as the 'magic number'.

a Copy this 3-by-3 square grid.
Place all the integers from −6 to 2 in your grid to make a magic square.

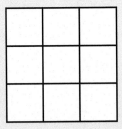

b Write down the magic number for your square.

9 A school hires two coaches for a trip to a museum. The first coach can carry 53 people and the second can carry 73 people. How many people can the school take to the museum?

10 Rashid has 277 fence panels on a truck. His first customer asks for 53 panels. His second customer wants 230. Does Rashid have enough panels on his truck? Use a clear numerical method to explain your answer.

11 A sports venue can hold 345 spectators. How many empty seats will there be if 283 spectators arrive?

12 A gym had 34 members in attendance on Wednesday, 24 on Thursday, 22 on Friday, 67 on Saturday and 49 on Sunday. How many more members attended the gym at the weekend (Saturday and Sunday) than on the other days?

13 Alysha is asked to subtract −27 from −8. She writes −35.
Explain why she is incorrect. Calculate the correct answer.

14 Write a calculation to check 600 − 247 = 353

15 Copy and complete each calculation.

Challenge Q15

a 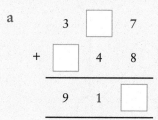   b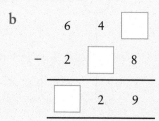

16 a Use the integers −7, 1, 3 and 9 to complete the calculation.

Challenge Q16

☐ + ☐ = ☐ − ☐

  b Can you find more than one way of answering part a?

17 Priti is thinking of two integers.
The difference between the two integers is 25.
The sum of the two integers is −1.
What two integers is Priti thinking of?

Challenge Q17

18 A three-year breeding program aims to breed and release 200 orangutans into their natural habitat.
During the first year of the program, 35 orangutans are released.
During the second year of the program, 52 orangutans are released.
During the third year of the program, 86 orangutans are released.
The target number of 200 is subtracted from the total number of orangutans released to give an answer of −27. What does this answer mean?

Challenge Q18

Investigation 1.1

Collect magazine, newspaper or online articles that use global temperatures, sea levels and carbon dioxide emissions. Explain in each case what the data tells you.

Research temperatures and the amount of rainfall for five different locations on the same day or month each year for 20 years. What does your data show you?

Investigation 1.2

Difference squares

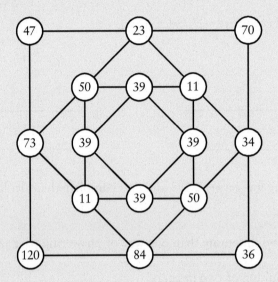

The diagram shows squares that are rotated by 45°.

There is a natural number at each of the four corners of the largest square. The positive difference of the two numbers at the end of each side of the square is written between them. For example, the positive difference between 47 and 70 is 23.

This process of finding the positive difference is applied to every square until all four differences are the same.

Draw your own difference square, choosing your own four numbers. Do you always have four squares in your diagram before all four differences are the same?

What is the least number of squares you can have? What is the most number of squares you can have?

Does this work for other shapes?

1.3.2 Multiplication and division

Multiplication and division are inverse operations of each other. This means that the act of dividing undoes the act of multiplying. For example, multiplying 4 by 5 gives 20. Dividing by 5 brings you back to 4 again. There is one number that this does not work for. Which number?

⑨ Explore 1.5

Work out 328 × 47. Share your method with others in the class. How many different methods have been used?

Here are two completed multiplication calculations using the Japanese method.

12 × 13   602 × 13

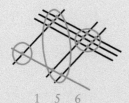   

    1   5   6   7   8   2   6

Can you explain how this method works? How does this method compare with the methods used by you or others in the class?

♀ Worked example 1.5

a   Write the answer to 23 ÷ 4 as a mixed number.

b   Use long division to work out 756 ÷ 23. Write the answer as a mixed number.

Solution

a $4{\overline{\smash{\big)}\,23}}$ $\overset{5\,r\,3}{}$ since 4 × 5 = 20, then the remainder $r = 3$ and thus the answer is $5\frac{3}{4}$

b $23{\overline{\smash{\big)}\,756}}$ $\overset{32}{}$

           −69↓   ←(3 × 23)
           66
           −46   ←(2 × 23)
           20   ←r

The answer is $32\frac{20}{23}$

🛡 Hint

This is one way of showing long division.

Some scientific calculators give the answer as a mixed number. You can use your calculator to check your answer.

dv

Reflect

Look back at the different methods from Explore 1.5. Are some methods more efficient or quicker than others?

Make a list of all the different words you can find that mean multiplication. Make another list for all the words you can find that mean division.

 Practice questions 1.3.2

1 Work out these questions.

a	$25 \div 5$	b	$16 \div 4$	c	$32 \div 4$
d	$18 \div 3$	e	$27 \div 3$	f	$36 \div 4$
g	$36 \div 6$	h	$45 \div 5$	i	$14 \div 7$
j	$21 \div 3$	k	$48 \div 6$	l	$40 \div 8$
m	$49 \div 7$	n	$56 \div 8$	o	$63 \div 7$
p	$72 \div 8$	q	$42 \div 7$	r	$45 \div 9$
s	$81 \div 9$	t	$72 \div 8$	u	$48 \div 8$

2 Copy and complete this multiplication table.

×	3	6	0	1	9	5		7	2	8		4
4							40				48	
					54							
7												
		48										

3 Show that:

a $24 \times 13 = 312$ b $73 \times 14 = 1022$

c $256 \times 15 = 3840$ d $324 \times 18 = 5832$

e $625 \times 25 = 15\,625$ f $426 \times 34 = 14\,484$

4   Work out each of the following. Write each answer as a mixed number.

   a $12 \div 5$ b $11 \div 4$ c $33 \div 10$

   d $45 \div 8$ e $7\overline{)12}$ f $4\overline{)5}$

   g $5\overline{)19}$ h $\dfrac{32}{3}$ i $\dfrac{11}{4}$

   j $\dfrac{58}{5}$ k $4\overline{)75}$ l $10\overline{)119}$

5   Work out each of the following. Write any remainders as fractions.

   a $10\overline{)3019}$ b $10\overline{)25\,000}$ c $10\overline{)12\,577}$

   d $4\overline{)597}$ e $5\overline{)1059}$ f $9\overline{)1089}$

   g $6\overline{)3648}$ h $6\overline{)36\,048}$ i $8\overline{)12\,573}$

6   Use long division to work out each calculation. Write any remainders as fractions.

   a $75 \div 12$ b $84 \div 14$ c $104 \div 13$

   d $153 \div 17$ e $199 \div 17$ f $205 \div 18$

   g $280 \div 21$ h $536 \div 23$ i $857 \div 16$

7   A school wants to take 367 people to a museum.

   Minibuses carry 16 people.

   How many minibuses will be needed?

8   A bottle of medicine contains 280 ml.

   Prita must take 15 ml of the medicine each day.

   For how many days will the medicine last?

   Give your answer as a mixed number.

9   Emil has 527 bags of cement on his truck.

   He sells them for $13 each.

   Show that Emil will earn $6851 if he sells all 527 bags.

10 Kiera orders 183 boxes of printer paper.

   Boxes come either as a single box for $23 or a pack of 8 boxes for $170.

   Find the cheapest way to order exactly 183 boxes.

   Show your mathematical method.

Challenge Q11

11 Copy and complete the calculation.

$$
\begin{array}{cccc}
 & 5 & 4 & 9 \\
\times & & & \boxed{} \\
\hline
\boxed{} & \boxed{} & 9 & 2 \\
\hline
\end{array}
$$

Challenge Q12

12 Arrange the digits 5, 6, 7 and 8 as a division calculation so that there is one digit in each box.

    a   What is the greatest possible answer?

    b   What is the least possible answer?

Challenge Q13

13 Use the calculation $34 \times 127 = 4318$ to write the answers to:

    a 34×1270 **b** 3400×127

    c $4318 \div 127$ **d** 68×127

    e $431\,800 \div 340$ **f** $8636 \div 127$

    g 17×127 **h** $8636 \div 17$

Thinking skills   🔍 **Investigation 1.3**

Draw a 2-by-2 grid and number it as shown.

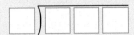

Trace your grid onto paper that you can see through, so you now have two copies of your grid.

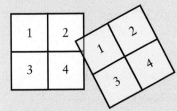

Place your traced grid over the top of your other grid so it fits exactly on the top.

You can turn the traced grid over or rotate it. The numbers do not have to match.

Multiply the numbers that are paired together and add your four answers together.

What is your total?

What is the highest possible total?

Try this for a 3-by-3 grid. Does the same orientation of the traced grid produce the highest total when multiplying the pairs of numbers together?

Does this work for any sized square grid?

How do you get the lowest possible total?

1.4 Order of operations

1.4.1 Order of operations without brackets

 Explore 1.6

Can you work out the answer to $4 + 5 \times 8 - 10 \div 2$?

What is the correct order of operations?

Work out the divisions and multiplications in a calculation before working out the additions and subtractions. If a calculation has only divisions and multiplications in it, work from left to right. If a calculation has only additions and subtractions in it, work from left to right.

 Worked example 1.6

Work out:

a $10 + 4 \times 3$ **b** $12 - 5 + 8$

Solution

a $10 + 4 \times 3 = 10 + 12$ ———— Work out 4×3 first.

$= 22$ ———— Then add 10 and 12.

A scientific calculator is programmed to work out calculations in the correct order. If you type $10 + 4 \times 3$ into a scientific calculator, it will give the answer 22.

b $12 - 5 + 8 = 7 + 8$ ⸻ The calculation involves only addition and subtraction, so work from left to right, one operation at a time.

$\qquad = 15$

 Reflect

Why is it important to have an order for operations?

Practice questions 1.4.1

1   Work out:

a	$23 + 14 + 7 + 2$	**b**	$37 + 18 + 12 + 9 + 3$
c	$30 - 12 - 7 - 2$	**d**	$16 - 7 - 4 - 3$
e	$9 + 8 - 5 + -2$	**f**	$11 + 7 - 8 - 6$
g	$28 - 11 + -9 - 4$	**h**	$17 - 14 + 5 - 10$
i	$42 - 21 - 9 + 7 - 3 + -5$	**j**	$12 - 2 - 7 + 4 - (-1) - 9$

2   Calculate:

a	$10 \times 5 \times 2 \times 2$	**b**	$2 \times 3 \times 3 \times 2$
c	$10 \times 2 \times 2 \times 2$	**d**	$3 \times 5 \times 2 \times 2$
e	$80 \div 20 \div 2$	**f**	$160 \div 4 \div 10 \div 2$
g	$180 \div 20 \div 3$	**h**	$8 \times 4 \div 2 \div 2$
i	$8 \times 6 \div 2 \div 3$	**j**	$100 \div 5 \div 4 \times 3$

3   Work out:

a	$10 - 2 \times 3$	**b**	$40 - 3 \times 3 \times 2$
c	$10 \times 2 - 2 \times 9$	**d**	$40 \div 5 + 12 \div 3$
e	$80 \div 20 - 6 \div 2$	**f**	$5 \times 8 + 6 \times 3$
g	$100 - 50 \div 2 - 20$	**h**	$125 - 25 \times 5$

Challenge Q4

4   Work out:

a $6 + 3 \times 7 - 25 \div 5 \times 3 + 9$

b $50 - 40 \div 5 + 6 \times 2 \div 4$

c $100 - 10 \times 8 \div 5 - 120 \div 6$

d $81 + 27 \div 3 - 10 \times 8 + 30 \div 3 \div 2$

5   Insert = or ≠ between each pair of calculations to make a correct statement.

Hint Q5

≠ means not equal to

   a $5 + 6 \times 8$ and $6 \times 8 + 5$

   b $5 + 6 \times 8$ and $5 \times 6 + 8$

   c $5 - 6 \times 8$ and $6 \times 8 - 5$

6   Copy and complete each calculation.

Challenge Q6

   a $3 + \boxed{} \times 5 = 23$

   b $4 \times \boxed{} - 12 \div 4 = 25$

Hint Q6d

The number missing from both boxes in part d is the same number.

   c $6 - \boxed{} \div 3 + 6 \times 7 = 38$

   d $8 + \boxed{} \times 4 - \boxed{} \div 2 = 50$

1.4.2 Order of operations with brackets

Explore 1.7

Can you work out $10 + 4 \times 5$ and $(10 + 4) \times 5$?

Are the answers the same?

What are the differences between the two expressions?

What general rules did you use in working out the calculations?

There are different types of brackets used to group numbers, calculations, expressions, letters and symbols in mathematics. The three most commonly used are parentheses ( ), brackets [ ] and braces { }.

Worked example 1.7

a   Work out $4 \times [20 - (3 + 9)]$

b   Work out $\dfrac{38 + 7}{15 - 6}$

Solution

a $4 \times [20 - (3 + 9)] = 4 \times [20 - 12]$ — First work out the calculation in the innermost bracket.

   $= 4 \times 8$ — Work out the calculation in the brackets.

   $= 32$

Hint

To work out $4 \times [20 - (3 + 9)]$ using a calculator, you need to use parentheses for both ( ) and [ ].

Hint

To work out $\dfrac{38 + 7}{15 - 6}$ using a calculator, type in $(38 + 7) \div (15 - 6)$ so that the calculator knows to work out the answer to the numerator and denominator first.

b $\dfrac{38 + 7}{15 - 6} = \dfrac{45}{9}$ ——— Work out the calculation in the numerator and the calculation in the denominator.

$= 5$ ——— Work out $45 \div 9$t

 Reflect

Write the rules for the order of operations. In what situations could you use your knowledge of the order of operations?

 Practice questions 1.4.2

1 Work out:

    a $(4 + 5) \times 10$ **b** $(15 - 5) \div 2$ **c** $7 \times (2 + 3)$

    d $9 \div (8 - 5)$ **e** $(100 - 5) \times 9$ **f** $20 \div [2 - (-3)]$

    g $20 - 3 \times 6$ **h** $(20 - 3) \times 6$ **i** $73 \times (46 - 45)$

2 Calculate:

    a $(4 + 7) \times (15 - 7)$ **b** $(9 - 5) \times (8 + 4)$ **c** $(32 + 4) - 6 \times 5$

    d $(11 + 9) \div (7 - 2)$ **e** $(10 - 4) \times (13 - 3)$ **f** $7 \times 4 \div (7 - 5)$

    g $100 - (5 + 3) \times 6$ **h** $88 \div 11 - (6 + 1)$ **i** $50 + (15 - 7) \div 2$

3 Work out:

    a $4 \times [7 + (10 - 7)]$ **b** $[5 + (8 - 4)] \times 9$

    c $100 - [10 + (6 - 5)]$ **d** $[(11 + 9) \div (7 - 2)] \div 2$

    e $100 - [(11 - 4) \times 11 - 3]$ **f** $(7 \times 4) \div [7 - (12 - 7)]$

    g $[100 - (5 + 3)] \times 2$ **h** $200 \div [28 - (30 - 22)]$

4 Work out:

    a $\dfrac{4 + 8}{3}$ **b** $\dfrac{31 + 4}{10 - 3}$ **c** $\dfrac{19 + 9}{21 \div 3}$

    d $\dfrac{4 \times 12}{4 + 4}$ **e** $\dfrac{42 \div 6}{15 - 8}$ **f** $100 - \dfrac{100 - 15}{45 \div 9}$

5   Some of these calculations give the same answer even when the brackets are removed. Which ones? In each case, explain why the brackets are needed or not needed.

Challenge Q5

a $3 + (4 \times 7)$

b $(5 + 6) \times 8$

c $(12 \times 7) + 24 \div 2$

d $(12 \times 7) + (24 \div 2)$

e $12 \times (4 + 6) \div 2$

f $(20 - 4) \times (10 - 7)$

6   Copy and complete each calculation.

Challenge Q6

a $(3 + \boxed{}) \times 5 = 40$

b $4 \times (\boxed{} - 10) \div 3 = 24$

Hint Q6d

The number missing from both boxes in part d is the same number.

c $(20 - \boxed{}) \times (5 + 6) \div 7 = 22$

d $(5 + \boxed{}) \times (10 - \boxed{}) \div 2 = 28$

7   Insert parentheses to make each of the following statements true.

Challenge Q7

a $7 + 8 \times 2 = 30$

b $4 \times 4 + 7 = 44$

c $30 \div 3 + 2 = 6$

d $4 + 2 \times 3 + 7 = 60$

e $4 + 2 \times 3 + 7 = 17$

f $4 + 2 \times 3 + 7 = 25$

g $100 - 5 + 3 \times 6 = 52$

h $75 \div 5 \div 5 = 75$

8   Use the numbers 2, 3, 5, 6 and 9 to make the calculation correct.

Challenge Q8

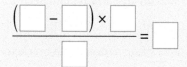

9   Use the numbers 1, 2, 3, 4, all four operations and parentheses to make the numbers 1 to 25.

Challenge Q9

Investigation 1.4

Thinking skills

Copy this table.

+	15	23	21	10
7				
12			33	
5				
8				

Complete the table by adding the numbers in the row and column headings. For example, 12 + 21 = 33

Circle any one of your answers and then cross out all of the other answers in that row and column. Circle a different answer that is not yet crossed out, then cross out all of the other answers in that row and column. Do this two more times, so that four of the answers are circled and the rest are crossed out.

Add the four circled numbers together. Work out the total of the eight numbers in the row and column headings. What do you notice?

Does the same happen when you change the starting numbers?

Copy this blank table.

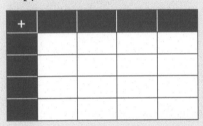

Choose your own eight starting numbers. Work through the same process as above. What do you notice? Explain why you think this happens.

Self assessment

I can identify natural numbers, integers and real numbers.

I can identify and use the place value of digits in natural numbers up to hundreds of millions.

I can identify and use the place value of digits in decimals.

I can add and subtract integers.

I can multiply and divide natural numbers.

I can use the order of operations without brackets.

I can use the order of operations with brackets.

? Check your knowledge questions

1 State the natural number that is 10 less than 321.

2 State the integer that is 20 less than 3.

3 Look at this list of numbers:

$-3, 0.5, 22, \frac{2}{3}, -10, -2.5$

From the list write down the:

a natural numbers   b integers   c real numbers.

4   Write down a real number between:

   a $\frac{1}{3}$ and $\frac{2}{3}$ b   −16 and −15   c   0.3 and 0.35.

5   Write down each of the numbers given in words as numerals.

   a   one thousand and twenty-seven

   b   six hundred and thirty-eight thousand, four hundred and ninety-six

   c   five hundred and four thousand, two hundred and one

   d   three million and eight.

6   Write down each numeral in words.

   a   3876   b   52 307   c   487 031   d   65 890 314

7   Write down the value of the 2 and the 6 in each number.

   a   2650   b   327 506   c   2 635 897

   d   0.206   e   23.5678   f   1.362

8   Write each group of numbers in ascending order.

   a   23 476, 2348, 230 687, 2 376 400   b   0.312, 0.3, 0.33, 0.321

9   Work out:

   a   2357 + 789   b   1562 − 985

10 Work out:

   a   −3 + 9   b   7 − 15   c   −8 − 13

   d   −35 + 7   e   −14 + −18   f   −23 − (−15)

11 Show that 57 × 248 = 14 136

12 Work out each calculation. Give your answer as a mixed number.

   a   40 ÷ 7   b $3\overline{)20}$ c $\frac{50}{8}$

13 Work out each calculation. Give your answer as a mixed number.

   a $8\overline{)5037}$ b $11\overline{)24\,000}$

14 Use long division to work out each calculation.
    Write any remainders as fractions.

   a   651 ÷ 26   b   865 ÷ 32

15 Use the calculation $231 \times 84 = 19\,404$ to write the answers to:

  a 84×2310 b 231×42 c $19\,404 \div 84$

  d 231×840 e $1\,940\,400 \div 231$

16 Work out:

  a $30 + 17 - 25 + (-6)$ b $20 - 9 - 7 - 12$

  c $15 - 3 - 8 + 6 - (-4) - 7$

17 Calculate:

  a $3 \times 4 \times 5 \times 6$ b $100 \div 5 \div 4$ c $12 \times 8 \div 2 \div 3$

18 Work out:

  a $50 - 4 \times 3 \times 2$ b $8 \times 3 - 24 \div 4$ c $40 - 60 \div 5 - 16$

  d $64 - 16 \times 4$ e $10 + 4 \times 9 - 30 \div 6 \times 4 + 7$

19 Calculate:

  a $(6 + 3) \times 11$ b $(23 - 5) \div 6$

  c $72 \div (5 - (-7))$ d $31 \times (25 - 21)$

20 Calculate:

  a $(1 + 10) \times (17 - 9)$ b $(25 + 35) \div (20 - 8)$

  c $100 - (5 + 3) \times 6$

21 Work out:

  a $12 \times [16 + (12 - 5)]$ b $[(15 + 9) \div (10 - 7)] \div 2$

22 Work out:

  a $\dfrac{72}{7 + 5}$ b $\dfrac{42 + 14}{15 - 8}$ c $30 - \dfrac{100 - 36}{24 \div 6}$

23 Some of these calculations give the same answer even when the brackets are removed. Which ones? In each case, explain why the brackets are needed or not needed.

  a $(5 \times 6) + (45 \div 9)$ b $8 \times (3 + 8) \div 4$

  c $(12 - 5) \times (20 - 10)$

24 Copy and complete the calculation $(7 \times \boxed{}) - 5 = 37$

Working
mathematically

2

2 Working mathematically

 KEY CONCEPT

Logic

 RELATED CONCEPTS

Models, Patterns, Representation

 GLOBAL CONTEXT

Scientific and technical innovation

Statement of inquiry

Representing scientific problems with logical mathematical models enables us to solve complex problems using innovative methods.

Factual

- What are the four steps in Pólya's problem-solving process?

Conceptual

- Why can it be useful to use a standard strategy when solving problems?

Debateable

- Are some problem-solving strategies better than others?

Do you recall?

What problem-solving strategies do you know?

2.1 Problem-solving

Throughout history, humans have used problem-solving strategies to advance scientific knowledge and create technological solutions to problems. Some major scientific and technological innovations include the printing press, electricity, vaccinations and the internet.

In and out of school, you will be presented with many problems to solve. In school, you might have to figure out how a chemical reaction works or how to convert a fraction to a decimal. Outside school, you might have to figure out the quickest route to get from your home to school, or what items to pack in a 10 kg bag for a seven-day trip. Either way, you will need to be able to solve a wide variety of problems. Knowledge of different problem-solving strategies will help you.

In this chapter, you will learn about different problem-solving strategies and how to use them to solve problems.

Problem-solving skills and thinking skills go hand-in-hand.

Choosing the right clothes to take on a trip is one problem you might have to solve, and working out the best way to fit them in your suitcase is another.

2.1.1 Pólya's four-step problem-solving process

In this series, we will use a four-step, problem-solving process outlined by George Pólya. The four steps are explained below.

Step 1: Understand the problem (UTP)

Read and re-read the problem carefully so that you understand it. Check your understanding by answering the following questions.

- Can you state the problem in your own words?
- What are you trying to find or do?
- What information is important?
- Is there any information provided that is not actually needed to solve the problem?
- Is any information missing?

Step 2: Make a plan (MAP)

Choose a strategy to solve the problem. Here is a list of strategies covered in this chapter:

- Trial and improvement
- Make a list
- Make a table
- Eliminate possibilities

> **Fact**
>
> George Pólya was a Hungarian mathematician (1887–1985). One of his biggest contributions to mathematics was his work on problem-solving. In 1945, he published the book *How to Solve It*, in which he outlined his systematic process for solving problems. This process is known as Pólya's four-step, problem-solving process.

> **Hint**
>
> Think about whether you have seen this problem or a similar one before.

- Make a diagram
- Look for a pattern
- Work backwards
- Simplify the problem.

The more problems you solve, the easier it will be to choose a strategy.

Step 3: Carry out the plan (COTP)

Carry out your plan to solve the problem. Make sure to check your work as you go. Lay out your work clearly so that others can understand and investigate the problem you have attempted.

Remember, you can use different strategies. Even if your original strategy does not work, it might lead you to a strategy that does.

Step 4: Look back (LB)

When you have finished working out the problem, you should look back on the work you have done.

- Did you answer the question that was asked?
- Does your answer make sense within the context of the question?
- Is your answer reasonable?

It is also important to reflect on the method you used.

- Could you have used another strategy to solve the problem?
- Will your method work for other problems?

💡 Worked example 2.1

Emma earns €2.00 every time she walks Chase, her neighbour's dog. Emma walks Chase twice a day, every day, for two weeks. She decides to save a quarter of the money she earns. How much money will Emma have saved from walking Chase by the end of the two weeks?

Solution

Understand the problem (UTP)

Sometimes it is helpful to highlight or underline the important information in a problem. If you want to, you can cross out any information that is not needed.

> Emma earns €2.00 every time she walks Chase, her neighbour's dog. Emma walks Chase twice a day, every day, for two weeks. She decides to save a quarter of the money she earns. How much money will Emma have saved from walking Chase by the end of the two weeks?

What is the important information?

- Emma earns €2.00, twice a day, every day, for two weeks. She saves a quarter of the amount she earns.

What are we asked to find?

- We need to calculate how much money Emma has *saved*.

Make a plan (MAP)

Think about what steps to follow, in order to calculate the answer.

We can start by working out what a quarter of €2.00 is. Then we will multiply by 2 to work out how much Emma saves every day. Finally, we will multiply by 14 to work out how much she saves in two weeks.

Carry out the plan (COTP)

$\frac{1}{4}$ of €2.00 = €0.50

$2 \times €0.50 = €1.00$

$14 \times €1.00 = €14.00$

Answer: Emma will have saved €14.00 by the end of the two weeks.

Look back (LB)

Emma walks the dog twice a day for two weeks so she earns €56 in total. That means that she could have saved between €0 and €56. As €14 is in this range, it is a reasonable answer.

 Reflect

Could we have used another strategy to solve the problem?

 Practice questions 2.1.1

1   Laurence is reading a book that has 182 pages. He read 33 pages on Monday night and 19 pages on Tuesday night. How many pages does he have left to read?

2   Mary has £90.00. If she buys a pair of jeans for £49.95 and a top for £17.50, how much money will she have left?

3   How many cars are needed to transport 62 children if each car can take 5 children?

4   Jeremiah lives 210 miles from his parents' house. After driving for three hours at 60 miles per hour, how many more miles does he need to drive to reach his parents' house?

5   Agnes cuts a pizza into eight equal slices. If she eats $\frac{1}{4}$ of the pizza, how many slices will be left?

6   Rudy bought 8 boxes of doughnuts, each containing 12 doughnuts. Each box cost ¥1000. Rudy then sold the doughnuts individually for ¥120 each. How much profit did he make?

 Hint Q6

Profit = selling price – cost

2.2 Strategies for problem-solving

 Explore 2.1

Can you solve the following problems?

a   Which option is better value?

Option 1: a pair of shoes costing €192 that you wear every day for two years

Option 2: a pair of shoes costing €135 that you wear every day for 1 year and 3 months

b   Each child in a family has at least two brothers and one sister. What is the smallest possible number of children in the family?

Being able to evaluate which products offer value for money is a useful problem-solving skill.

 Reflect

What strategies did you use to solve each problem? For example, did you draw a diagram? Make a table? Look for a pattern?

Discuss with someone else how they approached the problems. How are their approaches similar or different to yours?

Are some strategies better than others?

Were there any questions that you could not solve? If so, what did you learn from your approach to these problems?

As you might have discovered from Explore 2.1, the same problem can be solved using different strategies. Keep this in mind as you work through all the chapters in this series.

In the remainder of this chapter, you will learn about eight useful problem-solving strategies and how they work:

Trial and improvement

Guess ⟶ Check

Repeat ⟵ Improve

Make a list

1. _____
2. _____
3. _____

Make a table

	X	Y	Z
Boy			
Girl			

Eliminate possibilities

~~A.~~ ~~B.~~ C.

Make a diagram

Look for a pattern

Work backwards

Start at the end

Simplify the problem

P R O B L E M

2.2.1 Trial and improvement

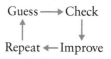

Guess ⟶ Check

Repeat ⟵ Improve

Explore 2.2

Ariadne and Vivienne are at a kitten café. They count the total number of legs and tails. After counting 18 more legs than tails, Ariadne challenges Vivienne to guess how many kittens are in the café. Using this information, Vivienne initially guesses that there are 3 kittens in the café.

She decides that if 3 is incorrect, her next guess will be 4 and if that does not work out, she will guess 5, and so on, until she works out the correct solution.

Discuss:

- How many kittens are there in the café?
- What strategy did Vivienne use to solve the problem?
- How could Vivienne have made a better first guess?
- Would you have solved the problem using the strategy Vivienne chose?

For the trial and improvement strategy, you will guess a possible solution and then work out the answer to see if your solution is correct. If your guess is incorrect, you will improve your guess and repeat the working. The process is repeated until you work out the correct solution. Trial and improvement is also known as *trial and error* and *guess, check and refine*.

 Worked example 2.2

A zookeeper counts the legs and heads of the penguins and elephants in the zoo. She counts 52 legs and 19 heads in total. How many elephants are there?

Solution

Understand the problem

What is the important information?

- There are 19 penguins and elephants all together.
- A penguin has two legs. An elephant has four legs. There are 52 legs in total.

What are we asked to find?

- We need to work out how many elephants there are.

Make a plan

Start by guessing how many elephants there could be. By subtracting our guess from 19, we can work out the corresponding number of penguins. We will work out how many legs there would be in total for our guess.

If the first guess does not give us the correct solution, we will use the answer to revise our guess up or down and repeat the process. We will continue to repeat the process until we work out a solution.

Carry out the plan

Guess 1

9 elephants

Check 1

If there are 9 elephants, then:

Number of penguins = 19 − 9 = 10

Work out the total number of legs: $(9 \times 4) + (10 \times 2) = 56$

This is too many legs!

Guess 2
7 elephants

Check 2
If there are 7 elephants, then:
Number of penguins = 19 − 7 = 12
Work out the total number of legs: $(7 \times 4) + (12 \times 2) = 52$
This is correct.

Answer
There are 7 elephants.

Look back

If we have 7 elephants and 12 penguins, we will have 19 animals altogether.
We can see from our working above that 7 elephants and 19 penguins have
a total of 52 legs.
Does the answer work? Yes.

 ## Practice questions 2.2.1

1   José is eight years older than Paul. The sum of their ages is 18.
    How old is José?

2   The sum of two numbers is 30 and their difference is 6.
    What are the numbers?

3   The product of two numbers is 35 and their sum is 12.
    What are the numbers?

4   A farmer has chickens and cows. If the chickens and cows have
    20 heads and 58 legs in total, how many chickens are there?

5   At a school bake sale, Dan sells cupcakes for 50 cents each and cookies
    for 30 cents each. He earns €6.60 in total. Twice as many cookies as
    cupcakes were sold. How many cupcakes were sold?

6   Margaret and Bo took their three children to the cinema. A child's
    ticket costs half as much as an adult ticket. The total cost of the tickets
    was £30.10. How much did each adult ticket cost?

7   Charlotte spent €3.50 purchasing four pencils and one ruler. A ruler
    costs 25 cents more than a pencil. How much does the ruler cost?

8   Ahmed, Luke and Josie collect yo-yos. Luke has twice as many as
    Ahmed. Josie has four more than Luke. They have 44 yo-yos in total.
    How many yo-yos does Josie have?

 Reminder

Operation	Symbol
Sum	+
Difference	−
Product	×
Quotient	÷

Challenge Q9

 Hint Q9

There is more than one solution.

 Challenge Q10

9 On the planet Zog there are two sorts of creatures.
   The Zigs have 3 legs and the Zags have 7 legs.
   Alana, the astronaut, saw a group of Zigs and Zags and counted
   that they had 61 legs between them.
   How many Zigs and how many Zags were there?

10 Michael is five years younger than Andre. Three times Michael's age
   plus Andre's age is equal to 33. What ages are Michael and Andre?

 Reflect

What are the advantages of the trial and improvement method?

What are the disadvantages of the trial and improvement method?

In what situations might trial and improvement not work at all?

In what situations might it take too much time?

2.2.2 Make a list

 Explore 2.3

Aisling wants to know how many 2-digit numbers contain a 7.

What strategy do you think Aisling should use to solve the problem? Why?

Aisling decides to make a list of all of the options: 17, 27, 37, …

By making a list, or using another method, count how many 2-digit numbers contain a 7.

Discuss: What advantages and disadvantages are there to making a list?

Making a list is a useful strategy for problems that have more than one solution, or when solving combination problems. It is important that you are organised and systematic when making your list to ensure that you account for all possible solutions.

 Worked example 2.3

Xun brings three T-shirts, two pairs of jeans and two hats on holiday.
How many unique outfits can he make?

Solution

Understand the problem

We need to work out how many different outfits Xun can put together.

Make a plan

We will use the following codes for the items of clothes:

- T-shirts: T_1, T_2 and T_3

- Jeans: J_1 and J_2

- Hats: H_1 and H_2

Make a list to organise the possibilities. In order to ensure we include all the possible outfits, we will list the options systematically. First, list the different combinations with T_1 before moving on to T_2 and T_3.

Carry out the plan

Possible outfits with T_1:

T_1, J_1, H_1 T_1, J_1, H_2 T_1, J_2, H_1 T_1, J_2, H_2

Possible outfits with T_2:

T_2, J_1, H_1 T_2, J_1, H_2 T_2, J_2, H_1 T_2, J_2, H_2

Possible outfits with T_3:

T_3, J_1, H_1 T_3, J_1, H_2 T_3, J_2, H_1 T_3, J_2, H_2

Answer

Xun can make 12 unique outfits.

Look back

With each T-shirt, Xun can make four unique outfits. As he has three T-shirts, and 3 lots of 4 is 12, we know the answer is correct.

Does the answer work? Yes.

 Reflect

Was it necessary to list all 12 of the possible outfits to work out the solution?

 Worked example 2.4

The numbers 1 to 10 are written on ten separate cards, one on each card.

a   How many pairs of cards have a sum of 8?

b   How many groups of three cards have a sum of 20?

Solution

Understand the problem

a   We need to identify pairs of cards that add to 8.
    How many pairs are there?

b   We need to identify groups of three cards that add to 20.
    How many groups are there?

Make a plan

Write out the numbers from 1 to 10.

a   List the pairs that add to 8. Be careful to use each number only once.

b   List the groups of three that add to 20. Be careful to use each number
    once only.

Carry out the plan

a   Start with 1 and work systematically, in order not to miss any solutions.

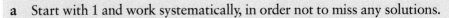

4 + 4 is not possible because 4 is used twice. 5 + 3, 6 + 2 and 7 + 1
have already been listed.

Answer
There are three pairs of cards that have a sum of 8.

b   As 20 is fairly big, we will start with 10 and see which pairs go with 10
    to make a sum of 20.

Now we will see which pairs go with 9. Remember that 9 + 1 + 10 has
already been listed, so we do not list it again here.

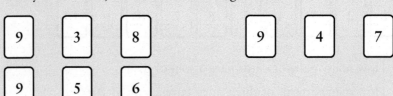

Now we will see which pairs go with 8. Remember that 8 + 2 + 10 and 8 + 3 + 9 have already been listed, so we do not list them again here.

| 8 | 5 | 7 |

8 + 4 + 8 and 8 + 6 + 6 are not possible because they use a number more than once.

If we see which pairs go with 7, we find that all of the options have already been listed.

No other groups of three cards will add to 20.

Answer
There are eight groups of three cards that have a sum of 20.

Look back

We have listed all of the possible outcomes systematically and not included any groups of cards twice or any groups that have a repeated number.

Does the answer work? Yes.

 Practice questions 2.2.2

1   Sabina is organising a trip to Europe. She would like to visit Paris, Berlin and London. If London must be the first or second city she visits, in how many ways can she organise her itinerary?

2   Three coins are tossed. List all the possible outcomes.

3   Amadi likes five books at the bookshop, but he only has enough money to buy two. From how many different pairs could he choose?

4   How many numbers less than 100 contain the digit 3?

5   If each digit is used only once, how many two-digit numbers can be made using the digits 3, 4, 5 and 6?

6   Venera owns a bookstore. In how many ways can she arrange three books, standing vertically, on a bookshelf?

7   Maria, Pedro and Adriana are being considered for Class A captain and vice-captain positions. Charlie and Julia are being considered for Class B captain and vice-captain. In how many different ways could the captains and vice-captains be chosen?

 Hint Q2

There are 8 possible outcomes.

 Hint Q3

Call the books A, B, C, D and E. The pair AB is the same pair of books as the pair BA.

 Challenge Q7

	X	Y	Z
Boy			
Girl			

2.2.3 Make a table

 Explore 2.4

Miss Jones asked all the students in her class to pick their favourite hobby from three options. She has the following results:

- 5 girls chose reading and 6 girls chose sport.

- 11 boys chose cooking.

- 24 students chose reading and 10 chose sport.

- There are 22 girls in the class.

Can you find out how many boys are in Miss Jones' class? How many of them chose reading?

Can you find out how many students chose cooking? How many of them are girls?

Did you use a table, and did it help you to solve the problem?

How would you solve the problem without using a table?

Some problems are easier to solve if you organise the information using a table. As you progress through the chapter, you will see that tables are a practical way to organise information when using other problem-solving strategies, too.

 Worked example 2.5

All of the adults and children at a party are drinking one of two available drinks: Orange Pop or Vanilla Cola. You have the following information:

- 16 adults are drinking Orange Pop.

- 25 people are drinking Vanilla Cola.

- There are 31 children at the party.

- 17 children are drinking Vanilla Cola.

How many people are at the party?

Solution

Understand the problem

We need to work out how many people are at the party.

Make a plan

Start by organising the information in a table. Then fill in any blanks to solve for the number of people at the party.

Carry out the plan

	Orange Pop	Vanilla Cola	Totals
Adult	16		
Child		17	31
Totals		25	

Filling in the blanks:

- Number of adults drinking Vanilla Cola: 25 − 17 = 8
- Number of children drinking Orange Pop: 31 − 17 = 14

Now we can fill these numbers into the table and work out the totals:

	Orange Pop	Vanilla Cola	Totals
Adult	16	8	24
Child	14	17	31
Totals	30	25	55

Answer
There are 55 people at the party.

Look back

The column total for Vanilla Cola adds up to 25. The row total for children adds to 31. Using this, we know that the column total for Orange Pop is 30 and the row total for adults is 24.

The totals for the rows and columns both add to 55.

Does the answer work? Yes.

 Practice questions 2.2.3

1 All of the students in Grades 5 and 6 eat lunch in the school canteen. One day, there are two main-course options: pizza and pasta. You have the following information:

- 22 Grade 5 students eat pasta.
- 13 Grade 6 students eat pizza.
- 39 students in total eat pizza.
- There are 35 students in Grade 6.

How many students in total are in Grades 5 and 6?

2   There are 96 people in Grades 1, 2 and 3. You have the following information:
- 37 children cannot swim.
- 11 children in Grade 1 cannot swim.
- 21 children in Grade 2 can swim.
- There are 30 children in Grade 3.
- 18 children in Grade 3 cannot swim.

How many children in Grade 1 can swim?

3   Penelope and Gigi go to the shop. They buy pencils, notebooks and highlighters. Pencils cost €0.50 each, notebooks cost €1.20 each and highlighters cost €0.75 each. You have the following information:
- Penelope and Gigi bought five pencils in total.
- Gigi bought three notebooks and Penelope bought two notebooks.
- Penelope bought two highlighters.
- Between them, Penelope and Gigi spent €5.25 on highlighters.
- Penelope spent €5.40 in total.

a   How many highlighters did Gigi buy?

b   How much did Penelope spend on pencils?

c   How much did Gigi spend in total?

d   What is the total amount of money spent by the two girls?

4   In a quiz, the teacher awards points depending on the difficulty of the question.
- A question = 10 points
- B question = 5 points
- C question = 2 points

Simon scores 29 points in the quiz. Copy and complete the table, showing different ways to obtain 29 points.

There is no limit to the number of questions that could be asked.

A question (10 points)	B question (5 points)	C question (2 points)	Total
1	1	7	29

5   There are 30 carpenters and bricklayers working on a building site.
    The workers travel to and from the site either by car, bus or bike.

* $\frac{1}{3}$ of the workers are carpenters.

*   Two carpenters go home by car.

* $\frac{4}{5}$ of the 10 people who go home by bike are bricklayers.

*   10 people go home by bus.

    How many carpenters go home by bus?

2.2.4 Eliminate possibilities

 Explore 2.5

Can you work out who Ben's teacher is using the information below?
Ben's teacher:

*   is not wearing a red top
*   is not wearing glasses
*   is female
*   does not have brown hair.

What strategy did you use to work out a solution?

Eliminating possibilities is a strategy in which you can work out the correct answers by eliminating answers that are incorrect. For many questions, it is helpful to start by organising the information in a table.

Worked example 2.6

Three friends are from different countries. Their last names are Smith, Garcia and Rossi. Their first names are Michelle, Olivia and Amalia, but not necessarily in that order. Smith is from England, Rossi has never been to Spain, Amalia is from Germany and Michelle is from Spain. Can you find out the full names of each of the friends?

Solution

Understand the problem

Using the information, we need to match the first names to the last names.

Make a plan

We can make a table to organise the solutions and then eliminate as many options as possible. From here, we will see if we can identify the names from the remaining information.

		England	Spain	Germany
Last name	Smith			
	Garcia			
	Rossi			
First name	Michelle			
	Olivia			
	Amalia			

Carry out the plan

Using the information given in the question, fill in the table as shown below:

		England	Spain	Germany
Last name	Smith	✓	✗	✗
	Garcia			
	Rossi		✗	
First name	Michelle	✗	✓	✗
	Olivia			
	Amalia	✗	✗	✓

We know that Smith and Rossi are not from Spain. That means that Garcia must be from Spain.

		England	Spain	Germany
Last name	Garcia	✘	✔	✘

We now know that Smith is from England and Garcia is from Spain, so Rossi must be from Germany.

		England	Spain	Germany
Last name	Rossi	✘	✘	✔

Similarly, we know Michelle is from Spain and Amalia is from Germany so Olivia must be from England.

		England	Spain	Germany
First name	Olivia	✔	✘	✘

Final completed table:

		England	Spain	Germany
Last name	Smith	✔	✘	✘
	Garcia	✘	✔	✘
	Rossi	✘	✘	✔
First name	Michelle	✘	✔	✘
	Olivia	✔	✘	✘
	Amalia	✘	✘	✔

Answer

The names are: Olivia Smith (from England), Michelle Garcia (from Spain) and Amalia Rossi (from Germany).

Look back

In the final solution, all of the friends have a unique name and they all come from different countries. Using the information given, there is no other solution possible.

Does the answer work? Yes.

 Practice questions 2.2.4

1   What odd number between 1 and 10 gives a remainder of 1 when divided by 3?

2	3	4	5	6	7	8	9

2   Cuddles is a zoo animal. You know the following information about Cuddles:
   - He does not live on land.
   - He is grey.
   - He has four legs.

Which animal is Cuddles?

3   What number between 10 and 40 has the sum of its digits equal to 7 and the difference of its digits equal to 3?

4   Olivia asked her aunt what age she was. Her aunt gave her three clues:
   - I am younger than 30.
   - The sum of the digits of my age is 9.
   - My age can be exactly divided by 6.

How old is Olivia's aunt?

5   A number of cards can be shared between three people exactly, but when shared between four people there are two cards left over and when shared between five people there are three cards left over. If there are fewer than 20 cards, how many cards are there?

6   If chocolate bars cost $1.30, $1.80 or $2.20, which of these could be the cost of three chocolate bars?
   A $3.80   **B** $8.90   **C** $4.80   **D** $5.60

7   Mila, William and Noah had breakfast together. Each chose a different item from eggs, cereal and fruit. You know the following information:
   - Mila sat next to the person who ate eggs.
   - William does not like cereal and is allergic to eggs.

What did each person eat for breakfast?

8 Brigid, Maura, Tom and Tristan are dressing up for a Hallowe'en party. Each of them is dressing as a different character. They will go as a skeleton, a witch, a pirate or a rabbit.
   - The pirate's name starts with T.
   - Maura is not the witch.
   - Brigid is not the skeleton.
   - Tristan is the rabbit.
   Who is dressing up as what character?

9 Cathal, Ryan and Nicola are friends. Their last names are Burns, Miller and Jones, but not necessarily in that order. Cathal lives in a red house, Nicola lives in a yellow house, the Burns live in a blue house and the Joneses do not live in a red house. What are the full names of each of the friends?

10 Ivan tells lies on Fridays, Saturdays and Sundays. He tells the truth on all other days. Robert tells lies on Tuesdays, Wednesdays and Thursdays. He tells the truth on all other days. If they both say, 'Yesterday, I lied,' what day is it today?

11 Darren, Cheng, Luke and Ali are friends. They each have a pet. Their pets are a turtle, a cat, a dog and a goldfish, but not in that order. You have the following information:
   - Luke plays basketball with the boy who has a cat.
   - Luke sings with the boy who has a dog.
   - Ali has been on holiday with the boy who has the turtle.
   - Ali travels to school with the boy who has the goldfish.
   - Darren is in the same class as the boy who has the turtle.
   - Darren is in the same class as the boy who has the cat.
   - Luke plays in a band with the boy who has the turtle.
   Which boy owns which pet?

12 Sasha, Marta, Zoé, Caterina and Lynn scored all of the points for their team in their school's basketball final. Skye, the school reporter, covered the game but she lost some of her notes. She knows the team's final score in the game was 95. Using the information from the notes she did not lose, work out how many points each player scored.

 Challenge Q12

   - Everybody scored an odd number of points.
   - Lynn scored 17 points. That was the fourth highest number of points.
   - Marta scored 12 more points than Sasha.
   - Sasha and Zoé scored a total of 30 points. Zoé scored more points than Sasha.
   - The last digit in everyone's score was different.
   - The highest score was 25 points.

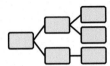

2.2.5 Make a diagram

Investigation 2.1

Coding is an important area of mathematics. For example, the RSA (Rivest–Shamir–Adelman) algorithm is used by computers and mobile phones to transmit messages securely over the internet.

A binary code, consisting of only 0s and 1s, can be used to send pictures. The image below shows a picture of Saturn made by the Voyager spacecraft on its tour of the outer planets between 1979 and 1989. This picture was communicated using the code shown on the left, which was then converted by computers into the image of Saturn shown on the below.

```
10011010110111100011001111001
01010010111010100100100000011
00011001111001100010111010101001
11100110001011101000110001011 1
00100110001010011000101 1101110
00110001011011010001101010001
00010111010010010111010100100 1
01001011101010010010000011010 1
11010100011011000101110111000 1
11001100010111010001111111 10011
```

How can we decode the following 77 digit message?

00010001000001110111000111111111000111111100000111110000000 01110
00000000100000

- The key to the code is that 0 represents white and 1 represents blue.
- The dimensions of a grid containing the 77 digits are prime numbers.

Because 77 is the product of the prime numbers 7 and 11, we know we need a grid that has dimensions of 11 × 7 or 7 × 11 in order to display the image.

We can then start working our way through the code from the top left to the bottom right. Copy and complete the picture for the code shown above.

Reminder

A prime number has only two factors: 1 and itself. The prime numbers less than 30 are 2, 3, 5, 7, 11, 13, 17, 19, 23, 29.

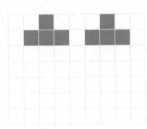

Reflect

Why do you think we use the product of two prime numbers to define the dimensions of the rectangle?

Making mathematical diagrams is a useful strategy that can help us to see problems more clearly and work out solutions.

 Explore 2.6

You are planning a party in a large recreation room. In order to seat all 22 party guests, you can use small tables that seat one person on each side. You need to arrange the tables so that they look like one large table.

Can you find the smallest number of tables to seat everyone?

Drawing a diagram is useful in many situations. You can use a diagram of your living space to work out how much carpet you need to buy, or if a new piece of furniture will fit.

 Worked example 2.7

Maya is building a fence around the perimeter of her garden. Her garden is rectangular in shape and is 12 metres long and 8 metres wide. Posts need to be placed two metres apart and there must be a post in each corner. How many posts does Maya need to buy?

Solution

We need to work out how many posts Maya needs to fence her garden.

Draw a diagram to see the problem more clearly. On the diagram, mark in the posts at two metre intervals so we can count how many are needed in total.

Since the length is 12 m and the space between posts must be 2 m, then there should be $\frac{12}{2} = 6$ spaces. Similarly with the width, there should be $\frac{8}{2} = 4$ spaces.

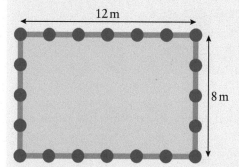

Maya needs 20 posts to fence her garden.

Looking back, every corner has a post and all posts are spaced 2 metres apart. The total number of posts shown in the diagram is 20.

Does the answer work? Yes.

💡 Worked example 2.8

Ling, Mohamed and Jack have a running race. Assuming a draw is not possible, in how many orders can they finish?

Solution

Understand the problem

A person can finish in 1st, 2nd or 3rd place. If a person comes in 1st place, then there are only two options left for 2nd place. Once you know who finished 1st and 2nd then there is only one choice left for 3rd place.

Make a plan

We can draw a diagram to organise the possible solutions. We will consider three different situations: when Ling finishes 1st, when Mohamed finishes 1st and when Jack finishes 1st.

Carry out the plan

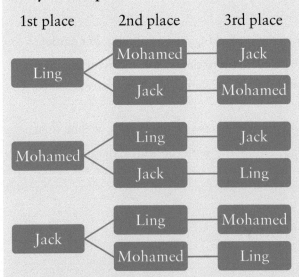

Answer

There are six ways that Ling, Mohamed and Jack can finish in a running race.

Look back

We can see from the diagram that all of the combinations have been listed. For example, if Ling finishes 1st, there are only two options left for 2nd and then one option left for 3rd, so there are only two ways Ling can come 1st. There are also only two ways Mohamed and Jack can come 1st.

So, 2 ways × 3 people = 6 possible outcomes.

Does the answer work? Yes.

What other strategies could you have used to work out the number of possible outcomes? Which strategy do you prefer? Why?

 Practice questions 2.2.5

1   Norah is building a block tower using four different coloured blocks. She places the red block below the green block. The blue block is placed above the yellow block, which is above the green block. What colour block is on top of her tower?

2   For his woodwork project, Harry has to hammer six nails into a piece of wood. The nails must be in a straight line and 1.5 cm apart. What is the distance from the first nail to the last nail?

3   A farmer is building a square pen for his chickens. He wants to use four posts for each side with one post in each corner. How many posts does he need in total?

4   A frog is at the bottom of an 11-metre well. Every day, it climbs up three metres and every night it slides back down one metre. If today is Monday, on what day will the frog reach the top of the well?

5   A spider is climbing up a 30-metre building. Each day, it climbs five metres and slides back one metre. How many days will it take to reach the top?

6   Five families are building houses in a large field. Paths will have to be built so that each house is directly connected to every other house. How many paths will have to be built?

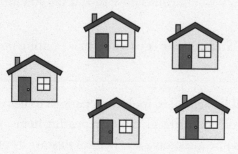

7   There are six people at a party. If every person at the party shakes hands with every other person there, how many handshakes will there be altogether?

8   A ball is dropped from a height of 8 metres. With each bounce, the ball reaches a height that is half the height of the previous bounce. After which bounce will the ball reach a maximum height of 25 cm?

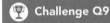 **Challenge Q9**

9   There are 22 students in Jeremy's class. Everyone eats one slice of pizza or one burger or both. Ten students eat a slice of pizza and four students eat both a slice of pizza and a burger. How many students eat a burger?

2.2.6 Look for a pattern

Explore 2.7

Luiz and Stefan are adding up the counting numbers from 1 to 10. They both choose a different strategy.

Luiz organises the numbers in ascending order and then adds them in order:

$1 + 2 + 3 + 4 + 5 + 6 + 7 + 8 + 9 + 10 = 55$

Stefan organises the numbers in pairs. He then finds the total of each pair and multiplies them to get an answer:

1, 2, 3, 4, 5, 6, 7, 8, 9, 10

$(1 + 10) + (2 + 9) + (3 + 8) + (4 + 7) + (5 + 6) = 5$ lots of 11
$$= 55$$

Both boys got the same answer.

Discuss:

- Which strategy is better?
- For Stefan's strategy, does it matter how he chooses his pairs? Why?
- Which strategy would you choose to add the counting numbers from 1 to 20?
- Can you use Stefan's strategy to add the counting numbers from 1 to 100?

 **Connections**

You will meet many problems involving patterns in Chapter 6, Patterns and rules.

Sometimes, it will be helpful to look for a pattern to help you to work out a solution to a problem. Patterns can be repeating items, shapes, images or numbers. For some questions, it is helpful to start by organising the information in a list or a table.

Worked example 2.9

Calculate:

a $4 + 3 + 2 + 4 + 3 + 2 + 4 + 3 + 2 + 4 + 3 + 2 + 4 + 3 + 2 + 4 + 3 + 2$

b $(1 + 2 + 3 + 4 + 5 + 6 + 7 + 8 + 9) + (9 + 8 + 7 + 6 + 5 + 4 + 3 + 2 + 1)$

Solution

Understand the problem

We need to calculate answers for the expressions shown.

Make a plan

We will look for patterns so we can group numbers and simplify the problem.

When we have simplified the problem, we will solve for the answer.

Carry out the plan

a The pattern '4 + 3 + 2' is repeating. By putting the numbers in groups of three, we can calculate a solution easily.

$(4 + 3 + 2) + (4 + 3 + 2) + (4 + 3 + 2) + (4 + 3 + 2) + (4 + 3 + 2) + (4 + 3 + 2)$

= 6 lots of 9

= 54

b If we add the first number from each group, we get 10. The second numbers also add to 10, and this pattern continues.

$$\begin{array}{r} (\ 1 + 2 + 3 + 4 + 5 + 6 + 7 + 8 + 9 \ ) \\ + (\ 9 + 8 + 7 + 6 + 5 + 4 + 3 + 2 + 1 \ ) \\ \hline 10 + 10 + 10 + 10 + 10 + 10 + 10 + 10 + 10 \end{array}$$

There are 9 lots of 10.

Answer
90

Look back

Do the answers work? Yes.

Double checking the working shows that the answers are correct.

Worked example 2.10

Maria is learning new words in Spanish. She learns two new words on the first day. Every day after that she learns three more words than on the previous day. How many Spanish words will Maria learn on day 10?

Solution

Understand the problem

We need to calculate how many words Maria learns on day 10. She adds 3 new words each day.

Make a plan

We will draw a table to organise the information and then look for a pattern.

Carry out the plan

Day	1	2	3	4	5	...	10
New words	2	5	8	11	14	...	

Looking at the pattern, the number of words learned on day 10:

= Number of words learned on day 1 + 9 lots of 3

$= 2 + 9 \times 3$

$= 29$

Answer

Maria will learn 29 new words on day 10.

Look back

Adding 3 new words every day for 10 days would make a total of 30 words on day 10. As she only learns two words on the first day, 29 new words on day 10 is a reasonable answer.

Does the answer work? Yes.

Practice questions 2.2.6

1 Calculate:

a $11 - 3 + 11 - 3 + 11 - 3 + 11 - 3 + 11 - 3 + 11 - 3 + 11 - 3 + 11$

b $5 + 6 - 4 + 5 + 6 - 4 + 5 + 6 - 4 + 5 + 6 - 4 + 5 + 6 - 4$

c $\frac{1}{2} - \frac{1}{3} + \frac{1}{3} - \frac{1}{4} + \frac{1}{4} - \frac{1}{5} + \frac{1}{5} - \frac{1}{6} + \frac{1}{6} - \frac{1}{7} + \frac{1}{7} - \frac{1}{8} + \frac{1}{8}$

d $50 - 49 + 48 - 47 + 46 - 45 + 44 - 43 + 42 - 41 + 40 - 39 + 38 - 37$

2 On Monday (day 1), Lily has €6 in savings. Starting from Tuesday, she saves €2 every day.

Calculate how much money Lily will have on:

a day 5   b day 10   c day 20.

3 George is making patterns with coins, blocks and pencils.

a

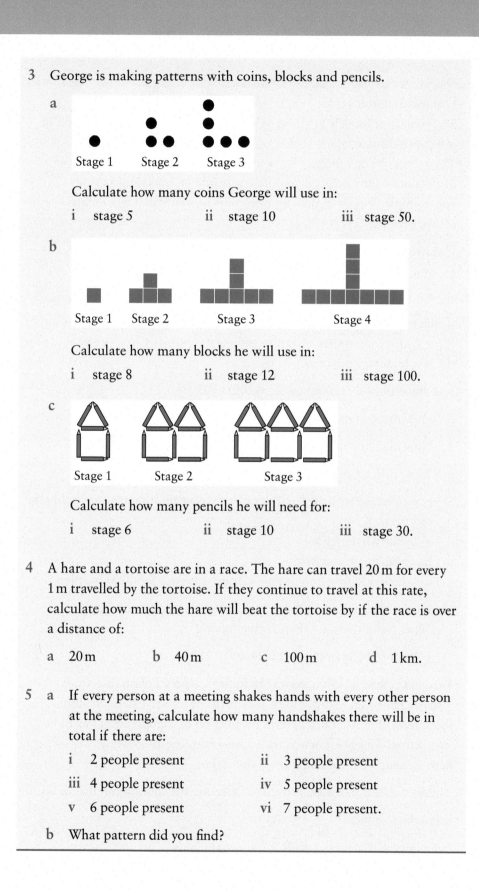

Calculate how many coins George will use in:

i   stage 5   ii   stage 10   iii   stage 50.

b

Calculate how many blocks he will use in:

i   stage 8   ii   stage 12   iii   stage 100.

c

Calculate how many pencils he will need for:

i   stage 6   ii   stage 10   iii   stage 30.

4 A hare and a tortoise are in a race. The hare can travel 20 m for every
1 m travelled by the tortoise. If they continue to travel at this rate,
calculate how much the hare will beat the tortoise by if the race is over
a distance of:

a   20 m   b   40 m   c   100 m   d   1 km.

5 a   If every person at a meeting shakes hands with every other person
at the meeting, calculate how many handshakes there will be in
total if there are:

i   2 people present   ii   3 people present

iii   4 people present   iv   5 people present

v   6 people present   vi   7 people present.

b   What pattern did you find?

Investigation 2.2

Disease dynamics

Mathematical models are used to make predictions about the spread of infectious diseases.

The reproduction number, R_0, is used to indicate the average number of people an infectious person infects during an infectious period. For example, if $R_0 = 2$, then an infectious person will infect an average of two people.

Questions:

1   What can you deduce about the spread of a disease if:
    a $R_0 < 1$ b $R_0 > 1$?

If one person in your class becomes infected with a disease that has an R_0 of 2, this means that each infected person infects two more people until the whole class is infected.

Assuming nobody is immune or vaccinated:

2   How many stages will it take to infect the whole class?

3   How many people are infected at each stage?

4   What difference would it make if the disease had an R_0 of 3?

5   For a disease with an R_0 of 2, estimate how many stages it would take to infect:
    a   your whole school
    b   your home country
    c   the world (the population is approximately 8 billion people).

6   How accurate do you think your solutions to question 5 are? What limitations are there to the model?

7   How could you improve the model?

Vaccines can be an effective way to limit the spread of an infectious disease. However, in order for a vaccine to be effective in controlling the spread of a disease, a certain proportion of a population has to be vaccinated. This is known as the vaccination threshold which will offer herd immunity to the remaining population.

The R_0 value of a disease is used to calculate the vaccination threshold using the formula: $\dfrac{R_0 - 1}{R_0} \times 100\%$

For example, the flu has an R_0 of approximately 2. So, the vaccine threshold for the flu is:

$$\frac{2-1}{2} \times 100\% = 50\%$$

This means that at least 50% of a population have to be vaccinated in order to control a future outbreak of the flu.

8 Research the R_0 values for at least four different diseases (e.g. measles, Ebola, chickenpox, common cold, polio, mumps, rubella, smallpox) and calculate the vaccination threshold for each disease.

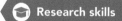

9 Research herd immunity. What problems could arise when a population relies on herd immunity?

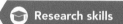

2.2.7 Work backwards

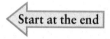

🌐 Explore 2.8

Calculate the number at the beginning of each number chain:

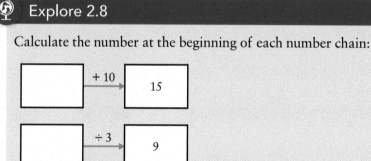

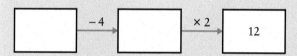

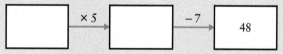

What strategy did you use to calculate the number at the beginning of each number chain?

How can you check that your answer is correct?

For some problems, the final outcome is already known, but the input is unknown. In these situations, working backwards can help you to work out a solution.

Worked example 2.11

Megan is 13 years younger than Lucas. Ariana is half Megan's age. Oliver is 7 years older than Ariana. If Oliver is 28, how old is Lucas?

Solution

Understand the problem

We know that the ages of four people are related. Ariana is younger than Megan, who is younger than Lucas and Oliver. Oliver's age is given and he is older than Ariana. We need to find Lucas's age.

Make a plan

We can use Oliver's age to work backwards and solve for Lucas's age.

Carry out the plan

Oliver is 28.
We know Oliver is 7 years older than Ariana:
Ariana = 28 − 7 = 21
We know that Ariana is half Megan's age. That is, Megan's age is twice Ariana's age:
Megan = 21 × 2 = 42
We know that Megan is 13 years younger than Lucas:
Lucas = 42 + 13 = 55

Answer
Lucas is 55 years old.

Look back

Working forwards:
Megan's age: 55 − 13 = 42
Ariana's age: 42 ÷ 2 = 21
Oliver's age: 21 + 7 = 28. This is correct.
Does the answer work? Yes.

 Practice questions 2.2.7

1 Rihanna thinks of three different numbers and gives the following clues.

  a When you add 3 and multiply by 5, the result is 35.

  b When you divide by 2 and then subtract 3, the result is 27.

  c When you add 4, multiply by 3 and then subtract 9, the result is 30.

  What number is Rihanna thinking of in each case?

2   Anne is 9 years younger than Peter. Larry is 15 years older than Anne. If Larry is 27, how old is Peter?

3   I am 35 years old. I was married 8 years ago and graduated 3 years before that. What was my age when I graduated?

4   Four children are counting their marbles. Mark has 17 less than David, Hayley has twice as many as Mark, and Tony has 9 more than Hayley. If Tony has 23 marbles, how many does David have?

5   Niamh has €14 now but during the last week she bought five ice-creams at €1.20 each and 7 drinks at 80 cents each. How much money did she have one week ago?

6   Patrick has a bag of sweets. Every day he eats 5 sweets and gives 7 away. After he had eaten the last 5 sweets, he had eaten 40 sweets altogether. How many sweets did he have at the start?

7   Derek has a bag of blueberries. He eats 5 blueberries and then gives half of what he has left to Eva. Eva eats 8 blueberries and then gives half of what is left to Brandon. Brandon eats 3 blueberries and gives half of what is left to Ruby. Ruby gets 4 blueberries. How many blueberries did Derek have in the bag at the start?

🏆 **Challenge Q7**

2.2.8 Simplify the problem

🔍 Explore 2.9

A hare challenges a tortoise to a race. The hare can run 20 metres for every 3 metres travelled by the tortoise. If they continue at this rate, by how much will the hare beat the tortoise if they race for a distance of 1000 metres?

Ronaldo decides to start by organising the information in a table:

Hare	20	40	60	80			
Tortoise	3	6	9				

How can drawing a table and looking for patterns help Ronaldo to solve the problem?

If a problem seems complicated, it can be helpful to try to solve a simpler problem or to see if you can solve the problem using easier numbers.

Worked example 2.12

There are 150 people taking part in a race. Each person is assigned a number from 1 to 150. How many people will have the number 7 as part of their number?

Solution

Understand the problem

We need to work out how many numbers between 1 and 150 contain at least one 7.

Make a plan

We have seen similar problems when we learned the 'Make a List' strategy. We could use a list again or we can try to work out a solution using a faster method. For example, we could start by working out how many numbers between 1 and 9 contain a 7. Then work out how many numbers between 10 and 19 contain a 7. We would continue this pattern until we reach 150. We will organise the results in a table.

Carry out the plan

Number	7s in ones position	7s in tens position
1–9	One: 7	
10–19	One:17	
20–29	One: 27	
30–69	Four: 37, 47, 57 and 67	
70–79	One: 77	Nine: 70, 71, 72, 73, 74, 75, 76, 78 and 79
80–89	One: 87	
90+	Six: 97, 107, 117, 127, 137 and 147	

Number of people with a 7 in their number = 1 + 1 + 1 + 4 + 1 + 1 + 6 + 9
$$= 24$$

Answer

24 people will have a number that contains a 7.

Look back

There are 15 tens in 150. Each set of 10 contains one number with a 7 in the ones position. This makes 15 numbers that contain a 7 in the ones position. In addition, we need to count the numbers with a 7 in the tens position. There are 9 extra numbers that have a 7 in the tens position (we do not need to count 77 again). 15 + 9 = 24, so we know our answer is correct.

Does the answer work? Yes.

Worked example 2.13

If it takes 15 people 10 hours to paint half of a house, how long will it take 4 people to paint the other half of the house?

Solution

Understand the problem

We need to calculate how long it will take 4 people to paint half of a house.

Make a plan

Start by looking at the problem using simpler numbers. We need to calculate how long it would take for one person to paint half of the house and then divide by 4 to work out how long it would take 4 people.

Carry out the plan

Looking at the problem using easier numbers:

If it takes 2 people 10 hours to paint half of the house, it would take 1 person 20 hours (10 × 2 = 20). So, if it takes 1 person 20 hours, it would take 4 people 5 hours to paint half of the house (20 ÷ 4 = 5).

Applying this logic to the example given:

If it takes 15 people 10 hours to paint half of the house, it would take 1 person 150 hours (10 × 15 = 150). So, if it takes 1 person 150 hours, it would take 4 people 37.5 hours to paint half of the house (150 ÷ 4 = 37.5).

Answer

It will take 37.5 hours for 4 people to paint the other half of the house.

Look back

As 1 person would need 150 hours to paint the house, 37.5 hours is a reasonable answer for 4 people to paint the house.

Does the answer work? Yes.

 Reflect

What was the key to solving this problem?

Could you have solved the problem using a different strategy?

Write a problem that is similar to the problem shown in Worked example 2.13.

Practice questions 2.2.8

1 In how many of the numbers between 1 and 150 do the digits 5 or 2 appear?

Challenge Q2

2 How many numbers between 1 and 1000 contain at least one 8?

3 A woman and her son are walking across a field. For every 9 metres the woman travels, her son travels 2 metres. By how far will the woman be ahead of her son after 180 metres?

Challenge Q4

Hint Q4

Think about how many diagonals you can draw from the corner of a square, then a pentagon (5 sides), then a hexagon (6 sides), and so on until you identify a pattern.

4 How many diagonals can be drawn from one corner of a figure on a flat surface that has 100 sides?

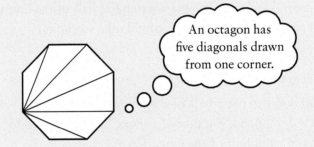

An octagon has five diagonals drawn from one corner.

Hint Q5

Think about what remainder you get when you multiply 5 × 5 and divide by 4, then the remainder you get when you multiply 5 × 5 × 5 and divide by 4, and so on.

5 If eleven 5s are multiplied together, what is the remainder when the answer is divided by 4?

6 A plane with 20 people on it runs out of fuel and has to land on a desert island. The 20 people have enough food and water to last for 12 days. Three days later, 4 more people in a sinking boat land on the island. How long will the rest of the food and water last now that there are 24 people in total?

 Self assessment

- I can apply Pólya's four-step problem-solving process when solving problems.
- I can use the following problem-solving strategies:
  - Trial and improvement
  - Make a list
  - Make a table
  - Eliminate possibilities
  - Make a diagram
  - Look for a pattern
  - Simplify the problem
- I can relate problems to similar problems that I have solved previously.
- I can show my working clearly and self-check for errors.
- I can reflect on my answers to ensure they make sense and are reasonable.

Kelly and Calvin are sister and brother. Answer the questions below using any strategy you like.

1. Kelly is three years older than Calvin. The sum of their ages is 19. How old are Kelly and Calvin?

2. Kelly wants to do chores around the house to earn money to save for a cinema trip with her friends in ten days' time. Her mother gives Kelly two options:

   Option 1: €2 every day for ten days.

   Option 2: €0.05 cents on the first day of the ten days, and double the amount from the previous day on each day after that.

   With which option will Kelly earn more money? How much money will she earn with this option?

3. After working and saving for ten days, Kelly goes to the cinema. At the cinema, she meets five friends. If each person shakes hands with every other person there, how many handshakes will there be altogether?

4. Kelly buys a ticket for the cinema and some snacks. She receives €5.50 change from €20.00. Which items did she buy?

5. In the theatre, the seats are organised into equal rows. When Kelly sits down, there are four seats to her left and four to her right. She sits in the fourth row from the front and the fifth row from the back. How many seats are there in the theatre?

6. Altogether, there are four girls and two boys. In how many ways can the six friends sit together if the boys sit together and the girls sit together?

Hint

If you are unsure of which method to use, think back to the problems you solved earlier in this chapter. If a problem has a similar structure to one you already know how to solve, the same strategy will probably work.

Hint Q6

$B_1B_2G_1G_2G_3G_4$ and $G_1G_2G_3G_4B_1B_2$ are two possibilities.

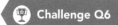

7 The film finished at 17:30. The friends sat down 2 hours and 20 minutes before the end, and the film started 25 minutes after they sat down. At what time did the film start?

8 At 18:00, after saying goodbye to her friends, Kelly decides to walk home. The distance between the cinema and Kelly's house is 1.2 km. At the same time Kelly starts walking, her mother leaves home to meet Kelly on the way. Kelly walks at a constant speed of 1 m/s and her mother walks at a constant speed of 1.5 m/s. At what time will Kelly meet her mother?

Number structure

13

3 Number structure

 KEY CONCEPT

Logic

 RELATED CONCEPTS

Equivalence, Patterns, Systems

 GLOBAL CONTEXT

Identities and relationships

Statement of inquiry

By understanding the relationships between real numbers, we can identify and use patterns to write them in equivalent forms.

Factual

• What is a prime number?

Conceptual

• How do you know if a common multiple is the lowest common multiple?

Debateable

• What different forms can numbers be written in?

Do you recall?

1 What are real numbers?

2 What written methods do you know for multiplying a 3-digit number by a 2-digit number?

3 How do you add and subtract fractions with the same denominator?

3.1 Number properties

There are various number properties that will be used in this chapter. Using these number properties can often make calculations easier to work out.

 Explore 3.1

What happens when you multiply a number by 1?

What happens when you multiply a number by 0 or add 0 to a number?

 Worked example 3.1

Work out 36 + 321 + 54 + 9

Solution

We need to rearrange the numbers to make the calculation easier to work out. Look for numbers to pair together that total multiples of 10 or 100 if possible.

$$36 + 321 + 54 + 9 = (36 + 54) + (321 + 9)$$
$$= 90 + 330$$
$$= 420$$

 Fact

When adding numbers, the sum is the same regardless of how the numbers are arranged. This is called the associative property.

 Reflect

Aisha, Badria and Dina work out 90 × 8 × 3

Aisha writes:
90 × 8 × 3 = 720 × 3 = 2160

Badria writes:
90 × 8 × 3 = 270 × 8 = 2160

Dina writes:
90 × 8 × 3 = 90 × 24 = 2160

Explain the method used by each person.

Do you think any of these methods are simpler than the others? If so, who do you think used the simplest method? Give a reason for your answer.

Practice questions 3.1

1 Work out the values of:

 a 5472×1
 b 459×0

 c $23 + 548$
 d $(2 \times 7) \times 4$

 e 1×572
 f $(14 + 81) + (5 + 27)$

 g $2 \times 2 \times 7$
 h $2 \times 7 \times 2$

 i $0 \times 1\,248\,457$
 j $5894 + 0$

2 Explain whether or not each equation is true.

 a $74 + 23 = 23 + 74$
 b $12 \times 15 = 15 \times 12$

 c $14 - 9 = 9 - 14$
 d $438 \times 1 = 438 + 0$

 e $712 \times 1 = 1 \times 712$
 f $25 \div 5 = 5 \div 25$

 g $63 - 8 = 8 - 63$
 h $647 + 0 = 0 + 647$

 i $647 - 0 = 0 - 647$
 j $647 \times 0 = 0 \times 647$

 k $100 \div 4 = 4 \div 100$
 l $236 + 1 = 236 \times 1$

3 For each equation, find the missing number that will make it true.

 a $\boxed{} \times 1 = 6871$
 b $\boxed{} + 0 = 472$

 c $339 + \boxed{} = 339$
 d $1643 \times \boxed{} = 0$

 e $\boxed{} + 19 = 19 + 49$
 f $6 \times 17 = \boxed{} \times 6$

 g $27 + 74 = \boxed{} + 27$
 h $472 \times \boxed{} = 472$

 i $8 \times \boxed{} = 6 \times 8$
 j $9 + \boxed{} = 97 + 9$

 k $124\,654 + \boxed{} = 124\,654$
 l $\boxed{} \times 11 = 11 \times 14$

4 Write down whether each statement is true or false.

 a $(12 + 9) + 7 = 12 + (9 + 7)$
 b $125 + (14 + 2) = (125 + 14) + 2$

 c $(32 \times 3) \times 5 = 32 \times (3 \times 5)$
 d $(24 - 8) - 4 = 24 - (8 - 4)$

 e $14 + 25 + 45 = 70 + 14$
 f $24 \div (4 \div 2) = (24 \div 4) \div 2$

 g $20 \times 5 \times 3 = 100 \times 3$
 h $91 + 14 + 9 = 14 + 100$

 i $36 \div 12 \div 3 = 36 \div 4$
 j $50 \times 5 \times 0 = 50 \times (5 \times 0)$

5   Write down the missing number that will make each number sentence true.

a $2 \times 8 \times 50 = \boxed{} \times 8$

b $476 + 14 + 16 = 476 + \boxed{}$

c $945 + 27 + 55 = \boxed{} + 27$

d $10 \times 257 \times 5 = 257 \times \boxed{}$

e $18 \times 12 \times 5 = \boxed{} \times 12$

f $97 + 1 = \boxed{} + 1 + 47$

6   Use what you have learned so far to find the answers quickly, without using a calculator.

a $5 \times 15 \times 2$

b $854 \times 15 \times 0$

c $30 + 148 + 70$

d $2457 + 600 + 400$

e $5 \times 4 \times 37 \times 5$

f $2 \times 19 \times 2 \times 25$

g $497 + 48 + 52$

h $72 + 45 + 28 + 55$

i $100 \times 50 \times 2 \times 10 \times 0$

j $700 + 597 + 3$

7   Work these out.

a $3 \times 4 \times 10 \times 10$

b $3 \times 40 \times 10$

c 3×400

d $30 \times 4 \times 10$

e 30×40

f 12×100

Why are the answers to a–f the same?

8   Write down whether each statement is true or false.

a   To multiply a number by 30, first multiply by 3 and then multiply by 10.

b   To multiply a number by 15, first multiply by 5 and then multiply by 3.

c   To divide a number by 12, first divide by 6 and then divide by 6 again.

d   To divide a number by 40, first divide by 4 and then divide by 10.

e   To multiply a number by 8, first double the number, then double the result and then double the result again.

f   To divide a number by 36, first divide by 6 and then divide by 6 again.

9   Use four different digits to complete each calculation.

a $\boxed{}\boxed{}\boxed{} \times \boxed{} = 0$

b $\boxed{}\boxed{}\boxed{} \times \boxed{} = 567$

10 Hamid multiplies by 21 using this method:

Multiply by 3 then multiply by 7.

For example, 47 × 21:

47 × 3 = 141

141 × 7 = 987

Explain why this method works.

11 Use Hamid's method from question 10 to work out these multiplications.

a   8 × 21   b   9 × 21   c   7 × 18

d   13 × 15   e   9 × 35   f   11 × 28

12 Nabila divides 690 by 15 using this method:

690 ÷ 3 = 230

230 ÷ 5 = 46

So, 690 ÷ 15 = 46

Explain why this method works.

13 Use Nabila's method from question 12 to work out these divisions.

a   480 ÷ 15   b   525 ÷ 15   c   540 ÷ 12

d   475 ÷ 25   e   315 ÷ 21   f   630 ÷ 35

 Challenge Q14

 Thinking skills

 Communication skills

14 Make a copy of this grid.

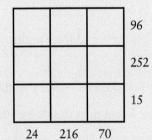

Use each of the digits 1 to 9 once to complete the grid. The digits in each row multiply to the answer next to that row, and the digits in each column multiply to the answer below that column.

3.2　Factors and multiples

A multiple of a number is found by multiplying the number by a positive integer. For example, the multiples of 3 are 3, 6, 9, 12, ... because $3 \times 1 = 3$, $3 \times 2 = 6$, $3 \times 3 = 9$, $3 \times 4 = 12$, and so on.

Explore 3.2

Can you find all of the numbers that divide exactly into 60? Write down your numbers in a list.

Can you write down three 4-digit numbers that are divisible by 2? Explain how you know that a number is divisible by 2.

 Hint

The numbers that divide exactly into 60 are factors of 60.

Although you can use a calculator to find factors of a number, it can often be quicker to use divisibility tests to work out factors, especially for larger numbers.

Explore 3.3

321 and 60 102 are both divisible by 3. Can you explain how you know that a number is divisible by 3?

Can you write down three 3-digit numbers that are divisible by 4? Can you explain how you know that a number is divisible by 4?

60, 125, 190 and 2135 are all divisible by 5. Can you explain how you know that a number is divisible by 5? Which of these numbers are divisible by 10? How do you know if a number is divisible by 10?

534 and 102 are both divisible by 6. Can you explain how you know that a number is divisible by 6?

81, 234 and 675 are all divisible by 9. Can you explain how you know that a number is divisible by 9?

60 has a lot of factors, which makes it useful for measuring time. The 60 minutes in an hour can be split evenly in many different ways.

Worked example 3.2

a　List all the factors of 84.

b　Find the highest common factor (HCF) of 84 and 140.

c　Find the lowest common multiple (LCM) of 8 and 12.

Solution

a　**Understand the problem**

We need to find all the numbers that divide exactly into 84.

Make a plan

Starting with 1, we will work systematically to find all the factor pairs of 84.

Carry out the plan

$1 \times 84 = 84$ 84 divides exactly by 1, so 1 and 84 are factors of 84

$2 \times 42 = 84$ 84 divides exactly by 2, so 2 and 42 are factors of 84

$3 \times 28 = 84$ 84 divides exactly by 3, so 3 and 28 are factors of 84

$4 \times 21 = 84$ 84 divides exactly by 4, so 4 and 21 are factors of 84

84 does not divide exactly by 5, so 5 is not a factor of 84

$6 \times 14 = 84$ 84 divides exactly by 6, so 6 and 14 are factors of 84

$7 \times 12 = 84$ 84 divides exactly by 7, so 7 and 12 are factors of 84

84 does not divide exactly by 8, 9, 10 or 11, so 8, 9, 10 and 11 are not factors of 84

We have already found a factor pair that includes 12, so we can stop here.

The factors of 84 are 1, 2, 3, 4, 6, 7, 12, 14, 21, 28, 42 and 84

Look back

Have we found all the factor pairs of 84? Yes, as $84 \div 10 = 8.4$. This answer is less than 10, so we know that there are no higher factors that we have not already listed.

b   We need to find the largest number that divides exactly into 84 and 140.

The plan is to list all the factors of 84 and list all the factors of 140. Then we will find the highest number that appears in both lists.

Factors of 84:   1, 2, 3, 4, 6, 7, 12, 14, 21, 28, 42, 84

Factors of 140:   1, 2, 4, 5, 7, 10, 14, 20, 28, 35, 70, 140

The common factors of 84 and 140 are the factors in both lists: 1, 2, 4, 7, 14, 28

The highest common factor of 84 and 140 is 28.

Is 28 the highest common factor of 84 and 140? Yes, 28 divides into 84 and 140 exactly. The only numbers higher than 28 that divide exactly into 84 are 42 and 84 but these do not divide into 140. The only numbers higher than 28 that divide exactly into 140 are 35, 70 and 140 but these do not divide into 84.

c We need to find the lowest number that is in both the 8 and 12 multiplication tables.

The plan is to list all the multiples of 8 and list all the multiples of 12. Then we will find the lowest number in both lists.

Multiples of 8:   8, 16, 24, 32, 40, 48, 56, …

Multiples of 12:   12, 24, 36, 48, 60, …

The lowest common multiple of 8 and 12 is 48.

Is 48 the lowest common multiple of 8 and 12? Yes, 48 is the lowest number in both the 8 and 12 multiplication tables. Extending the lists of multiples would show more common multiples, but the lowest common multiple is 48.

 Reflect

The divisibility tests from Explore 3.2 can be used to test the divisibility of other numbers. For example, if a number is divisible by both 3 and 4, it is also divisible by 12.

Use your results from Explore 3.2 to find other numbers that you can do divisibility tests for.

Can you find more common multiples in Worked example 3.2 part c without extending the lists?

 Practice questions 3.2

1 Write down the first six multiples of:

    a 4   **b** 11   **c** 2   **d** 10   **e** 5

2 Write down all the factors of:

    a 6   **b** 12   **c** 15   **d** 16   **e** 25

    f 100   **g** 30   **h** 40   **i** 27   **j** 99

3 Write down whether each statement is true or false.

    a 3 is a factor of 327   **b** 88 is a multiple of 8

    c 3 is a factor of 127   **d** 4 is a factor of 500

    e 333 is a multiple of 333   **f** 3 is a multiple of 333

    g 46 is a multiple of 6   **h** 1 is a factor of all natural numbers.

4   a   List the factors of 24.

    b   List the factors of 36.

    c   List the common factors of 24 and 36.

    d   Write down the highest common factor of 24 and 36.

5   Work out the highest common factor of:

    a   16 and 24   b   35 and 60

    c   70 and 100   d   35 and 63

    e   104 and 136   f   350 and 400

    g   144 and 168   h   49 and 25.

6   a   Write down the first ten multiples of 6.

    b   Write down the first ten multiples of 10.

    c   List the common multiples from parts a and b.

    d   Write down the lowest common multiple of 6 and 10.

7   Work out the lowest common multiple of:

    a   6 and 8   b   9 and 12

    c   25 and 15   d   10 and 8

    e   24 and 36   f   22 and 4

    g   14 and 21   h   12 and 15.

8   a   List the multiples of 4 between 50 and 70.

    b   List the multiples of 3 between 10 and 30.

    c   List the common factors of 120 and 150.

    d   Which multiples of 3, less than 100, are also multiples of 5?

    e   Find the multiples of 7 between 40 and 80.

    f   Which multiples of 7, less than 70, are also multiples of 3?

9   3, 4, 6, 10, 16, 25, 32, 40, 50, 60

    From this set of numbers, list the:

    a   factors of 20   b   multiples of 5

    c   factors of 100   d   multiples of 8.

10 Omar says, '10 is the lowest common multiple of 40 and 30.'
   He is incorrect.

   a   What mistake has Omar made?

   b   Write down the correct lowest common multiple of 40 and 30.

11 Write down two factors of 50 that add together to make 7.

Start by listing the factors
of 50.

12 What is the total of all the factors of 90 that are less than 10?

13 The sum of two factors of 80 is 26. Write down the two factors.

14 Write down a number that is:

   a   a factor of 24 and a factor of 30.

   b   a multiple of 5 and a factor of 40.

   c   a multiple of 6 and 8.

15 Ola writes down two numbers that have a highest common factor of
   15. What two numbers does Ola write down?

Start by listing numbers
with a factor of 15.

16 Olivia writes down two numbers that have a lowest common multiple
   of 15. What two numbers does Olivia write down?

17 The common factors of 30 and 36 are written in a list.
   Two of the numbers are selected and added together to get 9.
   What are the two numbers?

18 A teacher is making equipment packs for her students.
   She has 36 pens and 54 pencils.
   She wants each pack to be the same without any items left over.
   What is the greatest number of packs that she can make?

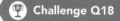

19 A street has three sets of traffic lights along its length.
   The first turns red every 60 seconds. The second turns red every
   90 seconds. The third turns red every 120 seconds.
   Khaled stands at the end of the street and sees all three sets of traffic
   lights turn red at the same time.
   How long must he wait for all three sets to turn red at the same time
   again?

3 Number structure

Challenge Q20

20 a Copy and complete this table.

	Multiples of 2	Factors of 18	Multiples of 5
Even numbers			
Odd numbers			
Multiples of 3			

b There is one cell in the table that cannot be completed. Explain why.

Challenge Q21

21 Investigate and write the divisibility test for:

   a 8   b 11   c 25

Thinking skills

 Investigation 3.1

Copy the 100 square grid.

1	2	3	4	5	6	7	8	9	10
11	12	13	14	15	16	17	18	19	20
21	22	23	24	25	26	27	28	29	30
31	32	33	34	35	36	37	38	39	40
41	42	43	44	45	46	47	48	49	50
51	52	53	54	55	56	57	58	59	60
61	62	63	64	65	66	67	68	69	70
71	72	73	74	75	76	77	78	79	80
81	82	83	84	85	86	87	88	89	90
91	92	93	94	95	96	97	98	99	100

Choose a number on the grid and cross it out. Now choose a multiple or factor of your number and cross that out. Keep repeating this process of choosing a multiple or factor of your previous number and crossing it out.

You can only use a number once.

Keep a record of your numbers in the order that you cross them out.

What is the longest chain of numbers that you can make?

Game: Fizz-buzz

This is a group game. Everyone in the group stands up, ideally arranged in a big circle, and one person is chosen to start.

This is a counting game that focuses on multiples of 3 and 5. The student starting the game says 1, the next student says 2, and so on. If the number on your turn is a multiple of 3, you say 'fizz' instead of the number. If the number on your turn is a multiple of 5, you say 'buzz' instead of the number. If the number on your turn is a multiple of both 3 and 5, you say 'fizz-buzz' instead of the number. For example, if your number was 12, you would say fizz, if it was 25, you would say buzz, and if it was 30, you would say fizz-buzz.

When a student makes a mistake, they sit down. Mistakes are when students say fizz, buzz, fizz-buzz or numbers in the wrong place.

The last three students standing are the winners.

Try this game with a different pair of multiples.

 Investigation 3.2

Consecutive multiples

Find two consecutive numbers where the first number is a multiple of 2 and the second number is a multiple of 3.

Find a few different examples. What do you notice? Explain your findings.

Does this work when the first number is a multiple of 3 and the second is a multiple of 4?

Does this work when the first number is a multiple of 4 and the second is a multiple of 5?

Find three consecutive numbers where the first number is a multiple of 2, the second number is a multiple of 3 and the third number is a multiple of 4.

Find a few different examples. What do you notice? Explain your findings.

 Thinking skills

 Communication skills

 Reminder

Consecutive numbers are numbers that follow on from each other. For example, 10, 11, 12.

Does this work for the multiples of 3, 4 and 5? What about the multiples of 4, 5 and 6?

Can this be applied to the multiples of the four consecutive numbers 2, 3, 4 and 5? How about the multiples of any four consecutive numbers?

How about the multiples of any five consecutive numbers?

3.3 Indices

3.3.1 Powers of numbers

We use indices/powers when multiplying the same number more than once.

Explore 3.4

Here is a sequence of squares.

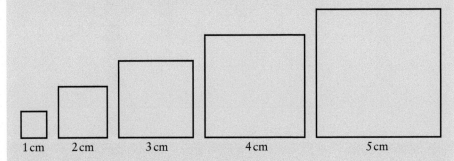

Can you work out the area of each square?

Can you carry on working out the area for squares up to side length 12 cm without drawing them?

Can you generalise your observations?

Here is a sequence of cubes.

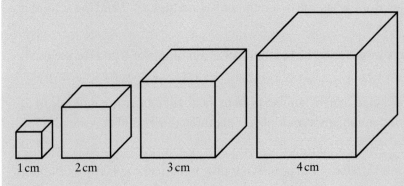

Can you work out the volume of each cube? Write down your calculation for each cube.

Can you carry on working out the volume of cubes up to side length 12 cm without drawing them?

The volume of each cube is a cube number. Can you explain how to work out cube numbers?

Worked example 3.3

a Write 6^4 in expanded form.

b Calculate 3^6

Solution

a $6^4 = 6 \times 6 \times 6 \times 6$ The index number or power is 4. This tells us how many times the base number, 6, is being multiplied together.

b $3^6 = 3 \times 3 \times 3 \times 3 \times 3 \times 3 = 729$ Calculators have a power button that may look like y^x or x^y.

Powers of numbers grow very quickly. If each square on a chess board was numbered $n = 1$ to 64, and 2^n grains of rice were placed on each square, there would not be enough rice in the world to finish the pattern.

 Reflect

Explain how you would work out the 5th square number or the 15th square number.

Explain how you would work out the 7th cube number or the 15th cube number?

Practice questions 3.3.1

1   Write each of these as a base number raised to a power.

a 4×4 b $7 \times 7 \times 7$

c $11 \times 11 \times 11 \times 11 \times 11$ d $25 \times 25 \times 25$

e $2 \times 2 \times 2 \times 2 \times 2 \times 2$ f $8 \times 8 \times 8 \times 8$

g $9 \times 9 \times 9 \times 9 \times 9$ h $12 \times 12 \times 12 \times 12$

2   Write each of these in expanded form.

a 4^3 b 5^4 c 10^6 d 7^5

e 3^6 f 6^7 g 15^4 h 9^5

3   Work out:

a 5^2 b 13^2 c 16^2 d 42^2 e 11^2 f 25^2

4   Calculate:

a 2^3 b 5^3 c 10^3 d 15^3 e 18^3 f 24^3

5   Work out:

a 2^7 b 3^5 c 6^4 d 5^5 e 12^5 f 12^4

6   Which is greater: 3^4 or 4^3?

7   Write down the value of:

a 2×10^1 b 4×10^3 c 7×10^5

d 3×10^4 e $2^2 \times 10^4$ f $3^2 \times 10^6$

Hint Q8

Work out powers/indices after brackets.

Connections

Order of operations, Chapter 1.

8   Write down the value of:

a $4^2 + 3^2$ b $5^2 \times 2^3$ c $8^2 \div 2^3$

d $(12 \div 4)^2$ e $(9 - 5)^3$ f $2^3 + 3^2 - 4^2$

g $12 + 2^3 \div 4$ h $(10 - 5)^2 + 3^3$ i $(3^3 - 4^2)^2 - 10^2$

9   Write down the missing numbers.

a $\boxed{}^2 = 16$ b $\boxed{}^2 = 36$ c $\boxed{}^2 = 81$ d $\boxed{}^3 = 8$

e $\boxed{}^3 = 125$ f $\boxed{}^3 = 27$ g $\boxed{}^3 = 1$ h $\boxed{}^2 = 64$

10 Salma has 70 square tiles. She arranges them in rows to make the largest square she can.

    a   How many tiles make up one edge of her square?

    b   How many tiles does Salma have left over?

11 Write down two square numbers that total:

    a   10   b   20   c   25

 Hint Q11

Start by listing square numbers.

12 Hamid has 100 square tiles.

He arranges them to make two different sized squares with no tiles left over.

How many tiles make up the edge of each of his squares?

13 Dina chooses two numbers less than 10.

She squares one number and cubes the other.

She then subtracts the square number from the cube number and the result is zero.

What two numbers did Dina choose?

 Hint Q13

Start by listing square and cube numbers.

14 Use the numbers 1, 2 and 3 to complete this calculation.

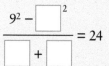

$$\frac{9^2 - \boxed{}^2}{\boxed{} + \boxed{}} = 24$$

15 Caria wants to work out $7^2 - 3^2$

She says, '$(7 - 3)^2$ gives me the same answer.'

Explain why Caria is incorrect.

16 Eymen has a box of 150 identical cubes.

He arranges them to make the largest cube possible.

    a   How many cubes has Eymen used?

Eymen wants to rearrange his cubes to build a larger cube.

    b   What is the minimum number of cubes Eymen needs to buy in order to build this larger cube?

17 Write the numbers 1, 2, 3, 15 and 34 into the boxes so that any pair of adjacent numbers adds to make a square number.

 Challenge Q17

 Thinking skills

3.3.2 Laws of indices

Explore 3.5

By writing 3^4 and 3^2 in expanded form, how can you write $3^4 \times 3^2$ as a single power?

Can you do the same with $3^3 \times 3^5$ and $5^4 \times 5^3$?

Compare the questions and answers. Can you find a quicker way to work out the answer?

Worked example 3.4

a Write $7^6 \times 7^2$ as a single power.   b Write $\dfrac{3^9}{3^5}$ as a single power.

Solution

a To write $7^6 \times 7^2$ as a single power, we need to express it as one single power of 7.

First expand each term, then combine them into one term.

$7^6 = 7 \times 7 \times 7 \times 7 \times 7 \times 7$ and $7^2 = 7 \times 7$, so:

$7^6 \times 7^2 = 7 \times 7 \times 7 \times 7 \times 7 \times 7 \times 7 \times 7$

$\qquad\qquad = 7^8$

b $3^9 = 3 \times 3 \times 3 \times 3 \times 3 \times 3 \times 3 \times 3 \times 3$

$3^5 = 3 \times 3 \times 3 \times 3 \times 3$

Therefore, $\dfrac{3^9}{3^5} = \dfrac{3 \times 3 \times 3 \times 3 \times 3 \times 3 \times 3 \times 3 \times 3}{3 \times 3 \times 3 \times 3 \times 3}$

As $\dfrac{3}{3} = 1$, $\dfrac{3^9}{3^5} = \dfrac{{}^1\cancel{3} \times {}^1\cancel{3} \times {}^1\cancel{3} \times {}^1\cancel{3} \times {}^1\cancel{3} \times 3 \times 3 \times 3 \times 3}{{}_1\cancel{3} \times {}_1\cancel{3} \times {}_1\cancel{3} \times {}_1\cancel{3} \times {}_1\cancel{3}} = 3^4$

Reflect

Try to write a rule for working out $\dfrac{n^a}{n^b}$

Practice questions 3.3.2

1 Copy and complete.

   a $2^3 \times 2^4 = 2^{\square}$ b $6^5 \times 6^3 = 6^{\square}$ c $7^2 \times 7^9 = 7^{\square}$

   d $9 \times 9^4 = 9^{\square}$ e $3 \times 3^1 = 3^{\square}$ f $4^3 \times 4^5 = 4^{\square}$

2   Write each of these as a single power.

   a $4^2 \times 4^3$ b $3^3 \times 3^2$ c $5^4 \times 5^7$ d $2^7 \times 2^5$

   e $8^{10} \times 8^3$ f $3^{11} \times 3^{11}$ g $10^7 \times 10^{14}$ h $12^1 \times 12^{12}$

3   Ayeleen writes $5^3 \times 5^4 = 5^{12}$
   Explain what she has done wrong.

4   Copy and complete.

   a $10^4 \times 10^{-1} = 10^{\square}$ b $10^2 \times 10^{-3} = 10^{\square}$

   c $10^5 \div 10^2 = 10^{\square}$ d $10^6 \div 10^4 = 10^{\square}$

   e $10^5 \div 10^{-3} = 10^{\square}$ f $\dfrac{10^7}{10^3} = 10^{\square}$

   g $\dfrac{10^2}{10^6} = 10^{\square}$ h $\dfrac{10^2}{10^{-3}} = 10^{\square}$

5   Write each calculation as a single power.

   a $4^4 \div 4^1$ b $4^8 \div 4^3$ c $5^3 \div 5^1$ d $6^7 \div 6^5$

   e $2^8 \div 2^5$ f $4^9 \div 4^{-4}$ g $4^7 \div 4^{-5}$ h $\dfrac{6^7}{6^3}$

   i $\dfrac{8^8}{8^2}$ j $\dfrac{5^7}{5^{-3}}$ k $\dfrac{7^2}{7^{-3}}$ l $\dfrac{2^{-2}}{2^{-5}}$

6   Write down:

   a   a multiplication calculation with the answer 6^{12}

   b   a multiplication calculation with the answer 4^{15}

   c   a division calculation with the answer 12^3

   d   a division calculation with the answer 5^8

   e   a division calculation with the answer 4^{-3}

   f   a division calculation with the answer 3^{-8}

   g   a multiplication calculation with the answer 4^{-4}

7   Copy and complete this multiplication grid.

 Challenge Q7

×	3^2	3^3	3^5		
3^{-3}					
	3^{-2}				3^4
3^6			3^{13}		
			3^{12}		
3^{-1}					

Challenge Q8

8 Copy and complete each calculation.

a $2^3 \times 2^\square = 2^9$ b $6^5 \times 6^\square = 6^{14}$ c $7^\square \times 7^9 = 7^{11}$

d $9 \times 9^\square = 9^3$ e $3^\square \times 3^{-1} = 3^7$ f $4^\square \times 4^{-3} = 4^6$

g $2^\square \div 2^3 = 2^7$ h $5^6 \div 5^\square = 5^2$ i $10^2 \div 10^\square = 10^{-2}$

 3.4 # Prime numbers and composite numbers

Explore 3.6

Try to list the numbers from 1 to 30 that are prime numbers.

Try to list the numbers from 1 to 30 that have more than two factors. These are called composite numbers.

Reminder

Numbers with exactly two factors are prime numbers.

Reminder

The word 'product' means multiply.

Prime numbers are crucial to keeping your details safe when you use online banking or shopping.

Hint

Recall problem-solving strategies from Chapter 2: Make a diagram!

Worked example 3.5

Use a factor tree to write 144 as a product of its prime factors.

Solution

We need to rewrite 144 as a multiplication calculation using its prime factors.

The plan is to draw a factor tree for 144 to find the prime factors of 144. Then we will use the factor tree to write 144 as a product of its prime factors.

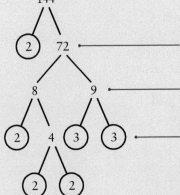

2 × 72 = 144, so write down 2 and 72. Circle the 2 as it is a prime number.

8 × 9 = 72, so write down 8 and 9. Neither 8 nor 9 are prime numbers so do not circle either of them but continue with both numbers.

2 × 4 = 8, so write down 2 and 4 branching off from the 8. 3 × 3 = 9, so write down 3 and 3 branching off from the 9. Circle the 2 and both 3s as they are prime numbers.

2 × 2 = 4, so write down 2 and 2. Circle both 2s as 2 is a prime number.

$144 = 2 \times 2 \times 2 \times 2 \times 3 \times 3$ ——— Use all of the prime numbers identified in the factor tree to write 144 as a product of its prime factors.

$\quad = 2^4 \times 3^2$ ——— Simplify the product of prime factors, if possible, using indices.

Is the product of prime factors for 144 correct? Yes, as 2 and 3 are both prime numbers, $2^4 = 16$, $3^2 = 9$ and $16 \times 9 = 144$

⇄ Reflect

Assume that 1 is a prime number whilst drawing the factor tree for 12. Explain what happens.

Can you explain why 1 is not a prime number or a composite number?

Amir, Imad and Nasser draw these factor trees.

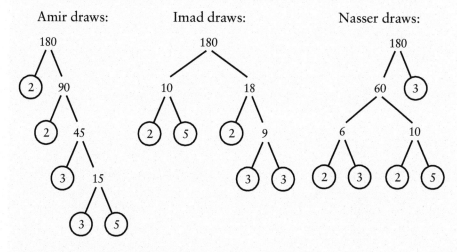

Amir draws: Imad draws: Nasser draws:

Explain why they are all correct.

Do you think any of the factor trees are easier or quicker to draw than the others? Give a reason for your answer.

Look back at Worked example 3.5. Can you find a different way to draw a factor tree for 144?

✏ Practice questions 3.4

1   a   How many odd prime numbers are there less than 30?

b   How many even prime numbers are there less than 30?

c   Write down the odd composite numbers that are less than 10.

d   Write down the composite numbers between 30 and 40.

2 Here is a list of numbers.

15, 3, 21, 2, 14, 20, 19, 25, 27, 4, 29, 33, 8, 9, 31, 12, 36, 13

   a   Write down the numbers from the list that are prime.

   b   Write down the numbers that are composite and odd.

3 For each statement, choose a number that makes it true.

   a   20 is a multiple of ☐   b   8 is a factor of ☐

   c   ☐ is a prime number between 20 and 30.

4 13 is a prime number.

Mateo says, 'The only common factor between 13 and any other number is always 1'.

Mateo is incorrect. Explain why.

5 Complete these factor trees.

   a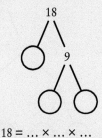

$$18 = \dots \times \dots \times \dots$$

   b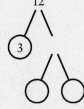

$$12 = \dots \times \dots \times \dots$$

   c

$$25 = \dots \times \dots$$

   d

$$30 = \dots \times \dots \times \dots$$

6 Write each number below as a product of its prime factors.

You must show your factor tree to display your method.

   a 20   b 50   c 45   d 75

   e 144   f 72   g 64   h 90

7 An alternative method to find the prime factors of a number is to continue to divide the number by prime numbers until you get a result of 1. Start with the lowest prime number that will divide into the number.

Use this method to write each of these numbers as products of their prime factors.

   a 180   b 225   c 252   d 1124

   e 512   f 1008   g 2700   h 3969

8   Write down two 2-digit prime numbers that total 30.

9   Xander draws a factor tree for the prime factor decomposition of 27.

Xander has made a mistake.

What is his mistake?

$27 = 3 \times 9$

10 Zikra draws this factor tree.

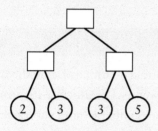

What number did Zikra start with?

🏆 **Challenge Q11**

11 6 is called a 'perfect' number because the factors of 6, other than 6 itself, total 6:

$1 + 2 + 3 = 6$

Find a perfect number between 20 and 30.

🔍 Investigation 3.3

Sieve of Eratosthenes

Copy the 100 square grid.

Cross out the number 1.

Circle the number 2 and then cross out all the remaining multiples of 2.

Circle the number 3 and then cross out all the remaining multiples of 3.

1	2	3	4	5	6	7	8	9	10
11	12	13	14	15	16	17	18	19	20
21	22	23	24	25	26	27	28	29	30
31	32	33	34	35	36	37	38	39	40
41	42	43	44	45	46	47	48	49	50
51	52	53	54	55	56	57	58	59	60
61	62	63	64	65	66	67	68	69	70
71	72	73	74	75	76	77	78	79	80
81	82	83	84	85	86	87	88	89	90
91	92	93	94	95	96	97	98	99	100

🔷 **Fact**

Eratosthenes was a Greek mathematician who was born in 276 BCE and died in 194 BCE. He was also a geographer, astronomer, poet and historian.

4 is already crossed out so the next number is 5. Circle the number 5 and then cross out the remaining multiples of 5.

Continue doing this until every number is either circled or crossed out.

What set do all of the circled numbers belong to?

Why was the number 1 crossed out?

Why do all the 2-digit prime numbers end in a 1, 3, 7 or 9?

How many prime numbers are there between 1 and 20? How many are there between 21 and 40, 41 and 60, 61 and 80 and 81 and 100? Write down any observations about your results.

Do you think the total number of prime numbers between 1 and 100 will be the same as between 101 and 200? Why do you think this? Are you correct?

 Thinking skills

 Research skills

Investigation 3.4

1   Choose a 2-digit prime number and subtract 1 from it. Draw the factor tree for your answer and write your answer as a product of its prime factors. For example, choosing the prime number 67 and subtracting 1 gives 66 and $66 = 2 \times 3 \times 11$. Working backwards, can you create a prime number? Do you always get a prime number?

2   Write down the highest common factor of each pair of numbers.

   a   8 and 13   **b**   2 and 9   **c**   13 and 25.

   What do you notice about your answers?

   Each pair of numbers is coprime. What do you think coprime means? Research coprime numbers to see if you are correct.

3   Choose two prime numbers. Work out the lowest common multiple of your two prime numbers.

   Try this for a few different pairs of prime numbers.

   What do you notice about your answers? Why does this happen?

 3.5   Fractions

3.5.1 Fractions and mixed numbers

A fraction has a numerator and a denominator. The numerator is the top number in the fraction and the denominator is the bottom number.

For example, $\frac{2}{3}$ has a numerator 2 and a denominator 3.

A mixed number is a number made up of a whole number and a fraction, for example, $2\frac{1}{4}$

Can you explain why $\frac{1}{3}, \frac{2}{6}, \frac{3}{9}$ and $\frac{4}{12}$ are equivalent fractions? What other fractions can you find that are equivalent to these fractions?

Can you find any fractions that are equivalent to $\frac{1}{4}, \frac{2}{3}$ and $\frac{4}{5}$? Explain your method.

What fractions can you find that are equivalent to $\frac{1}{4}, \frac{2}{3}$ and $\frac{1}{2}$ that all have the same denominator? Use your equivalent fractions to order the fractions $\frac{1}{4}, \frac{2}{3}$ and $\frac{1}{2}$ from smallest to largest.

Can you use the same method to write $\frac{3}{5}, \frac{1}{2}, \frac{5}{8}$ and $\frac{7}{10}$ in order from smallest to largest?

We are often asked to simplify answers that involve fractions.

Explore 3.8

Sayed uses this method to simplify $\frac{15}{25}$

Can you explain Sayed's method? Why has he divided the numerator and denominator by 5?

Can you use Sayed's method to simplify $\frac{16}{20}$? Can you explain how to simplify $\frac{16}{20}$? Which is the quickest method? Explain the quickest method to simplify a fraction where the numerator and denominator have more than one common factor greater than 1.

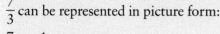
Write $\frac{7}{3}$ as a mixed number.

Solution

$\frac{7}{3}$ is an improper fraction because the numerator is greater than the denominator.

$\frac{7}{3}$ can be represented in picture form:

$\frac{7}{3} = 2\frac{1}{3}$

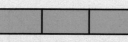
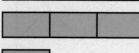

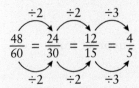 Reflect

How can you write $\frac{7}{3}$ as a mixed number without drawing a diagram?

Explain your method.

Find and explain a method for converting the other way, changing $2\frac{1}{3}$ to an improper fraction.

Sara simplifies $\frac{48}{60}$ like this:

$$\frac{48}{60} = \frac{24}{30} = \frac{12}{15} = \frac{4}{5}$$

Explain how Sara could have simplified $\frac{48}{60}$ using fewer steps.

 Thinking skills

 Investigation 3.5

Tangrams

A tangram is a puzzle that originated from China and is made up of seven pieces.

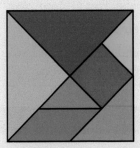

You can make your own tangram using squared paper like this:

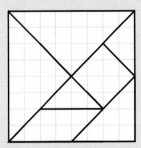

Copy and complete the table on the next page, writing the fraction that the left shape is of the top shape.

Shape from tangram				
	$\frac{1}{4}$			

Tangrams are used to make pictures. There are hundreds of different pictures that can be made. Here is a tiny sample:

Make your own tangram and create these pictures. What pictures of your own can you make?

Try to arrange all the pieces of the tangram in a different way to make the same sized square. Can it be done?

Is it possible to arrange the pieces to make a rectangle?

Practice questions 3.5.1

1 Copy and complete the equivalent fractions.

a $\frac{1}{2} = \frac{\square}{4} = \frac{5}{\square} = \frac{\square}{10} = \frac{8}{\square}$

b $\frac{1}{10} = \frac{\square}{40} = \frac{10}{\square}$

c $\frac{1}{5} = \frac{\square}{40} = \frac{5}{\square}$

d $\frac{2}{3} = \frac{\square}{6} = \frac{\square}{15}$

e $\frac{3}{4} = \frac{\square}{8} = \frac{9}{\square} = \frac{24}{\square}$

f $\frac{3}{2} = \frac{6}{\square} = \frac{\square}{12}$

2 Write down the fraction that is *not* equivalent to the other fractions in each list.

a $\dfrac{1}{2}, \dfrac{4}{8}, \dfrac{8}{14}, \dfrac{16}{32}$

b $\dfrac{4}{12}, \dfrac{5}{18}, \dfrac{3}{9}, \dfrac{1}{3}$

c $\dfrac{5}{20}, \dfrac{2}{5}, \dfrac{12}{30}, \dfrac{4}{10}$

d $\dfrac{5}{10}, \dfrac{1}{4}, \dfrac{10}{20}, \dfrac{16}{32}$

e $\dfrac{40}{30}, \dfrac{16}{12}, \dfrac{4}{3}, \dfrac{3}{4}$

f $\dfrac{20}{200}, \dfrac{10}{25}, \dfrac{40}{100}, \dfrac{2}{5}$

3 Sort these fractions into sets of equivalent fractions.

$\dfrac{1}{6}$ $\dfrac{5}{50}$ $\dfrac{4}{6}$ $\dfrac{10}{60}$ $\dfrac{44}{55}$ $\dfrac{2}{3}$ $\dfrac{16}{20}$ $\dfrac{50}{100}$ $\dfrac{100}{1000}$ $\dfrac{7}{42}$

$\dfrac{1}{10}$ $\dfrac{5}{30}$ $\dfrac{20}{30}$ $\dfrac{10}{100}$ $\dfrac{20}{25}$ $\dfrac{22}{33}$ $\dfrac{20}{200}$ $\dfrac{80}{100}$ $\dfrac{4}{5}$ $\dfrac{3}{18}$

Which fraction is the odd one out?

4 Simplify each fraction.

a $\dfrac{8}{10}$

b $\dfrac{10}{25}$

c $\dfrac{18}{36}$

d $\dfrac{12}{48}$

e $\dfrac{45}{100}$

f $\dfrac{28}{49}$

g $\dfrac{18}{72}$

h $\dfrac{28}{40}$

5 Which of these fractions cannot be simplified?

$\dfrac{6}{10}$ $\dfrac{15}{50}$ $\dfrac{35}{49}$ $\dfrac{13}{45}$ $\dfrac{81}{99}$ $\dfrac{33}{35}$ $\dfrac{89}{100}$ $\dfrac{85}{100}$ $\dfrac{15}{22}$

Explain why these fractions cannot be simplified.

6 Three students are asked to simplify $\dfrac{28}{40}$

Hamid writes:

Aref writes:

Mirac writes:

$$\dfrac{28}{40} \overset{\div 2}{=} \dfrac{14}{20}$$

Explain who is correct.

7 Place one of these signs, <, = or >, between each fraction pair to make the statement true.

a $\dfrac{1}{2}$ ☐ $\dfrac{13}{25}$

b $\dfrac{3}{10}$ ☐ $\dfrac{1}{4}$

c $\dfrac{7}{10}$ ☐ $\dfrac{35}{50}$

d $\dfrac{66}{100}$ ☐ $\dfrac{2}{3}$

e $\dfrac{18}{24}$ ☐ $\dfrac{24}{32}$

f $\dfrac{2}{9}$ ☐ $\dfrac{2}{10}$

g $\dfrac{8}{10}$ ☐ $\dfrac{8}{9}$

h $\dfrac{4}{5}$ ☐ $\dfrac{5}{7}$

8 Place these fractions in ascending order.

   a $\dfrac{1}{2}, \dfrac{1}{4}, \dfrac{4}{10}$
        b $\dfrac{3}{10}, \dfrac{2}{5}, \dfrac{1}{3}$
        c $\dfrac{5}{20}, \dfrac{2}{10}, \dfrac{30}{100}$

   d $\dfrac{2}{3}, \dfrac{6}{10}, \dfrac{67}{100}, \dfrac{11}{20}$
        e $\dfrac{1}{2}, \dfrac{8}{10}, \dfrac{3}{4}, \dfrac{13}{20}$

9 This list of fractions is in ascending order. Copy and fill in the missing numbers.

$$\dfrac{\square}{4}, \dfrac{1}{3}, \dfrac{\square}{2}, \dfrac{3}{\square}, \dfrac{3}{\square}, \dfrac{8}{10}$$

10 Aya has these numbered cards.

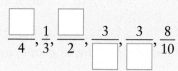

   She chooses two cards to make a fraction.

   a What fraction does she make to get a fraction greater than $\dfrac{1}{2}$?

   b What fraction does she make to get a fraction less than $\dfrac{1}{2}$?

11 Write down each improper fraction as a mixed number.

   a $\dfrac{5}{2}$
        b $\dfrac{19}{10}$
        c $\dfrac{14}{5}$

   d $\dfrac{45}{8}$
        e $\dfrac{55}{7}$
        f $\dfrac{349}{100}$

12 Write down each mixed number as an improper fraction.

   a $2\dfrac{1}{3}$
        b $1\dfrac{9}{10}$
        c $2\dfrac{2}{5}$

   d $4\dfrac{3}{11}$
        e $10\dfrac{6}{7}$
        f $21\dfrac{4}{7}$

13 Write down the fractional amount of each number.

   a $\dfrac{1}{2}$ of 30
    b $\dfrac{1}{3}$ of 30
    c $\dfrac{1}{5}$ of 30
    d $\dfrac{1}{6}$ of 30

   e $\dfrac{2}{3}$ of 30
    f $\dfrac{3}{5}$ of 30
    g $\dfrac{5}{6}$ of 30
    h $\dfrac{7}{10}$ of 30

14 Write down the fractional amount of each number.

   a $\dfrac{1}{4}$ of 20
    b $\dfrac{1}{5}$ of 100
    c $\dfrac{1}{8}$ of 32
    d $\dfrac{3}{5}$ of 100

   e $\dfrac{3}{4}$ of 24
    f $\dfrac{7}{8}$ of 16
    g $\dfrac{9}{10}$ of 1000
    h $\dfrac{9}{100}$ of 500

15 Work out each of these, giving each of your answers in smaller units if necessary.

a $\frac{1}{4}$ of \$10

b $\frac{1}{3}$ of \$1.20

c $\frac{3}{4}$ of 2.4 kg

d $\frac{2}{5}$ of 5.5 m

e $\frac{3}{8}$ of \$2.40

f $\frac{7}{10}$ of 1 hour.

16 a If $\frac{1}{2}$ of a bottle of water is 2340 ml, how much water will the bottle hold when full?

b If $\frac{1}{4}$ of a sports stadium holds 1200 people, how many people will fill the stadium?

c If $\frac{1}{10}$ of my time spent streaming films is 45 minutes, how long do I spend streaming films?

d If $\frac{3}{4}$ of an aquarium holds 27 litres, how much would the aquarium hold when $\frac{1}{2}$ full?

e If $\frac{3}{10}$ of a town's population is 7200, what is $\frac{9}{10}$ of the town's population?

17 Dalia says, 'I'm thinking of a number. $\frac{2}{3}$ of my number is 24.'
What is Dalia's number?

18 Place one of these signs, <, = or >, between each pair of calculations to make the statement true.

a $\frac{1}{2}$ of 32 ☐ $\frac{1}{4}$ of 44

b $\frac{1}{4}$ of 40 ☐ $\frac{1}{5}$ of 60

c $\frac{2}{3}$ of 24 ☐ $\frac{3}{4}$ of 20

d $\frac{1}{5}$ of 50 ☐ $\frac{2}{3}$ of 15

e $\frac{1}{6}$ of 120 ☐ $\frac{3}{8}$ of 56

f $\frac{9}{10}$ of 200 ☐ $\frac{3}{8}$ of 400

19 Nia received 120 text messages this week.

$\frac{1}{4}$ of the messages are from Nia's parents and $\frac{2}{5}$ are from her best friend.

How many of her messages are *not* from her parents or her best friend?

20 Felix and Hatem share their trading cards.

Felix gets $\frac{3}{8}$ of the cards, which is 45 cards.

How many cards does Hatem get?

21 There are 900 students at a school.

$\frac{3}{5}$ buy a hot meal from the school canteen at lunch time.

$\frac{1}{4}$ buy a cold meal from the school canteen at lunch time.

The rest bring their own food to eat for lunch.

How many students bring their own food for lunch?

22 Zahra receives an allowance each month from her parents.

As a special one-off reward, she is offered either $\frac{2}{3}$ extra on this month's allowance or $\frac{1}{10}$ of her annual allowance.

Which option gives Zahra the larger reward? Show your mathematical reasoning.

 Thinking skills

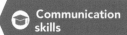 Communication skills

3.5.2 The four operations with fractions and mixed numbers

Explore 3.9

The diagram shows $\frac{2}{3}$ and $\frac{1}{4}$. How does the diagram show that $\frac{2}{3} + \frac{1}{4} = \frac{11}{12}$?

$$\frac{2}{3} \qquad \frac{1}{4}$$

Can you explain how to work out $\frac{2}{3} + \frac{1}{4}$ without using a diagram?

Can you use a similar method to work out $\frac{2}{3} - \frac{1}{4}$?

Explore 3.10

Here are three different methods for working out $\frac{2}{3} \times \frac{1}{4}$

Method 1: $\qquad \frac{1}{4}$ of $\frac{2}{3} = \frac{2}{12} = \frac{1}{6}$

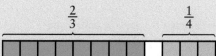

Method 2:

$$\frac{2}{3} \times \frac{1}{4} = \frac{2 \times 1}{3 \times 4} = \frac{2}{12} = \frac{1}{6}$$

Method 3:

$$\frac{2}{3} \times \frac{1}{4} = \frac{\cancel{2}^1 \times 1}{3 \times \cancel{4}_2} = \frac{1}{6}$$

Can you explain how each method works?

Can you use all three methods to work out $\frac{4}{5} \times \frac{5}{6}$?

Which method do you prefer? Explain why.

Worked example 3.7

Work out:

a $\frac{3}{4} \div \frac{2}{3}$ b $5\frac{1}{2} - 1\frac{2}{3}$ c $2\frac{3}{4} \times 1\frac{2}{3}$ d $3\frac{1}{2} \div 2\frac{4}{5}$

Solution

a $\frac{3}{4} \div \frac{2}{3} = \frac{3}{4} \times \frac{3}{2}$ ⟵ Make 1 the denominator like this: $\dfrac{3}{4} \div \dfrac{2}{3} = \dfrac{\frac{3}{4}}{\frac{2}{3}} = \dfrac{\frac{3}{4} \times \frac{3}{2}}{\frac{2}{3} \times \frac{3}{2}} = \dfrac{\frac{3}{4} \times \frac{3}{2}}{1}$

$\qquad = \frac{3 \times 3}{4 \times 2}$ You can use a calculator to check your answer by using the

$\qquad = \frac{9}{8}$ or $1\frac{1}{8}$ $\boxed{a\frac{b}{c}}$ or $\boxed{\frac{a}{b}}$ function.

b $5\frac{1}{2} - 1\frac{2}{3} = (5 - 1) + \left(\frac{1}{2} - \frac{2}{3}\right)$ ⟵ Subtract the whole numbers and fractions separately.

$\qquad = 4 + \left(\frac{3}{6} - \frac{4}{6}\right)$

$\qquad = 4 + \left(-\frac{1}{6}\right)$ ⟵ $4 + \left(-\frac{1}{6}\right) = 3\frac{6}{6} + \left(-\frac{1}{6}\right)$

$\qquad = 3\frac{5}{6}$

c $2\frac{3}{4} \times 1\frac{2}{3} = \frac{11}{4} \times \frac{5}{3}$ ⟵ First change both mixed numbers to improper fractions.

$\qquad = \frac{11 \times 5}{4 \times 3}$

$\qquad = \frac{55}{12}$ or $4\frac{7}{12}$

d $3\frac{1}{2} \div 2\frac{4}{5} = \frac{7}{2} \div \frac{14}{5}$ ⟵ First change both mixed numbers to improper fractions.

$\qquad = \frac{7}{2} \times \frac{5}{14}$ ⟵ Make 1 the denominator.

$\qquad = \frac{\cancel{7}^1 \times 5}{2 \times \cancel{14}_2}$

$\qquad = \frac{5}{4}$ or $1\frac{1}{4}$

Use your preferred method of multiplying fractions to work out $\dfrac{16}{75} \times \dfrac{25}{32}$

Is there a different method that you would prefer to use for larger numerators and denominators?

Do you know any alternative methods for the worked examples?

✎ **Practice questions 3.5.2**

1 Work out each calculation.

Give each answer in its simplest form.

a $\dfrac{1}{10} + \dfrac{1}{5}$ b $\dfrac{1}{2} + \dfrac{3}{10}$ c $\dfrac{2}{3} + \dfrac{1}{5}$ d $\dfrac{5}{8} + \dfrac{1}{2}$

e $\dfrac{1}{3} + \dfrac{1}{2}$ f $\dfrac{2}{3} + \dfrac{1}{4}$ g $\dfrac{1}{4} + \dfrac{3}{5}$ h $\dfrac{9}{10} + \dfrac{1}{3}$

2 Work out each calculation.

Give each answer in its simplest form.

a $\dfrac{7}{10} - \dfrac{1}{5}$ b $\dfrac{1}{2} - \dfrac{2}{10}$ c $\dfrac{2}{3} - \dfrac{1}{6}$ d $\dfrac{3}{5} - \dfrac{1}{2}$

e $\dfrac{7}{8} - \dfrac{2}{3}$ f $\dfrac{3}{4} - \dfrac{2}{3}$ g $\dfrac{9}{10} - \dfrac{2}{3}$ h $\dfrac{7}{8} - \dfrac{2}{5}$

3 Work out each calculation.

Give each answer in its simplest form.

a $\dfrac{13}{20} + \dfrac{2}{8}$ b $\dfrac{17}{25} + \dfrac{3}{10}$ c $\dfrac{1}{2} - \dfrac{6}{25}$ d $\dfrac{47}{100} + \dfrac{3}{5}$

e $\dfrac{2}{3} - \dfrac{3}{8}$ f $\dfrac{7}{12} - \dfrac{3}{8}$ g $\dfrac{9}{7} + \dfrac{3}{4}$ h $\dfrac{17}{20} - \dfrac{7}{25}$

4 Work out each calculation.

Give each answer in its simplest form.

a $1\dfrac{2}{3} + \dfrac{5}{6}$ b $2\dfrac{5}{8} + \dfrac{7}{10}$ c $3\dfrac{3}{5} + \dfrac{9}{10}$ d $1\dfrac{5}{8} + 2\dfrac{1}{3}$

e $5\dfrac{1}{2} + 4\dfrac{3}{7}$ f $10\dfrac{7}{20} + 2\dfrac{1}{25}$ g $7\dfrac{2}{5} + 12\dfrac{3}{4}$ h $3\dfrac{3}{5} + 7\dfrac{5}{6}$

5 Simplify:

a $1\dfrac{2}{3} - \dfrac{5}{6}$ b $5\dfrac{1}{2} - \dfrac{2}{3}$ c $2\dfrac{2}{5} - 1\dfrac{7}{10}$ d $4\dfrac{2}{3} - 1\dfrac{3}{4}$

e $8\dfrac{3}{10} - 7\dfrac{2}{3}$ f $12\dfrac{13}{100} - 5\dfrac{13}{20}$ g $5\dfrac{5}{12} - 3\dfrac{5}{8}$ h $3\dfrac{3}{100} - 1\dfrac{11}{40}$

6 Work out and simplify.

a $\dfrac{3}{5}$ of $\dfrac{3}{10}$ b $\dfrac{5}{6}$ of $\dfrac{4}{7}$ c $\dfrac{3}{5} \times \dfrac{4}{7}$ d $\dfrac{3}{8} \times \dfrac{4}{5}$

e $\dfrac{9}{12} \times \dfrac{4}{7}$ f $\dfrac{3}{10} \times \dfrac{5}{9}$ g $\dfrac{10}{12} \times \dfrac{8}{25}$ h $\dfrac{9}{20} \times \dfrac{5}{9}$

7 Work out and simplify.

a $\dfrac{3}{5} \div \dfrac{1}{2}$ b $\dfrac{1}{2} \div \dfrac{1}{5}$ c $\dfrac{9}{10} \div \dfrac{9}{10}$ d $\dfrac{3}{8} \div \dfrac{2}{5}$

e $\dfrac{10}{11} \div \dfrac{5}{8}$ f $\dfrac{1}{2} \div \dfrac{1}{50}$ g $\dfrac{35}{24} \div \dfrac{7}{12}$ h $\dfrac{4}{5} \div \dfrac{3}{4}$

8 Work out each calculation.

Give each answer in its simplest form.

a $1\dfrac{1}{3} \times \dfrac{3}{4}$ b $1\dfrac{2}{3} \times \dfrac{3}{5}$ c $\dfrac{3}{5} \times 2\dfrac{3}{4}$ d $\dfrac{5}{8} \times 3$

e $4 \times \dfrac{3}{10}$ f $2\dfrac{1}{5} \times 1\dfrac{1}{6}$ g $3\dfrac{2}{3} \times 5$ h $1\dfrac{2}{3} \times \dfrac{1}{4}$

9 Calculate and simplify.

a $4 \div \dfrac{1}{5}$ b $1\dfrac{3}{4} \div \dfrac{1}{5}$ c $4\dfrac{4}{5} \div \dfrac{2}{3}$ d $3\dfrac{4}{7} \div \dfrac{5}{8}$

e $2\dfrac{3}{5} \div 5$ f $3\dfrac{3}{4} \div 5$ g $3\dfrac{1}{8} \div 1\dfrac{1}{4}$ h $4\dfrac{2}{7} \div 3\dfrac{1}{3}$

10 Kira trains for Tae Kwon Do.

She works on her fitness for $\dfrac{1}{3}$ of an hour.

She then works on her kicks for $\dfrac{3}{10}$ of an hour.

What fraction of an hour does Kira spend training?

11 Ashraf is playing a board game.

He loses $\dfrac{3}{8}$ of his pieces in the first 10 minutes and $\dfrac{1}{3}$ in the second 10 minutes. The rest are still on the board at the end of the game.

a What fraction of his pieces has Ashraf lost in the game?

b What fraction of his pieces has he left at the end?

12 Copy and complete this calculation.

$$\dfrac{7}{12} + \dfrac{\boxed{}}{7} = 1\dfrac{13}{84}$$

13 Work out the area of the rectangle as a fraction of $1\,\text{m}^2$.

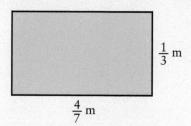

$\frac{1}{3}$ m

$\frac{4}{7}$ m

🛡 Hint Q13

Area of a rectangle = length × width

14 Hiba has a lunch break which is $\frac{9}{10}$ of an hour long.

She eats her lunch and then goes for a walk.

Hiba divides her time equally between eating her lunch and walking.

What fraction of an hour does she spend walking?

15 Mazin and Darien are training for a race.

Mazin has run $1\frac{2}{5}$ km and Darien has run $1\frac{1}{8}$ km.

How much farther than Darien has Mazin run?

16 For this rectangle, work out:

a   the perimeter

b   the area.

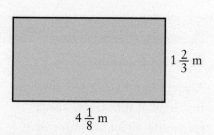

$1\frac{2}{3}$ m

$4\frac{1}{8}$ m

17 In this pyramid, each block in the upper layers is found by adding together the two numbers below it.

Copy and complete the number pyramid.

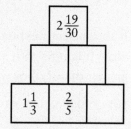

$2\frac{19}{30}$

$1\frac{1}{3}$ $\frac{2}{5}$

🏆 Challenge Q17

18 Write down three mixed numbers with different denominators that total 20.

🏆 Challenge Q18

19 Write down the missing number.

$2\frac{1}{4} + 3\frac{3}{8} - \boxed{} = 7\frac{1}{2}$

🏆 Challenge Q19

Challenge Q20

20 Copy and complete the multiplication grid.

×	$1\frac{1}{3}$		$2\frac{1}{4}$
$2\frac{2}{9}$			
$1\frac{3}{5}$		$1\frac{23}{25}$	
			$6\frac{3}{16}$

 Challenge Q21

21

$1\frac{1}{8}$ $\quad$ $\frac{2}{5}$ $\quad$ $\frac{2}{3}$ $\quad$ $1\frac{1}{3}$ $\quad$ $\frac{3}{10}$

Find a pair of cards that gives:

a the smallest total

b the greatest total

c the greatest difference

d the greatest product

e the smallest product

f the smallest positive difference.

Thinking skills

Investigation 3.6

Egyptian fractions

Ancient Egyptians used only unit fractions. A unit fraction is a fraction with a numerator of 1, for example, $\frac{1}{2}, \frac{1}{3}, \frac{1}{4}$, and so on.

If the ancient Egyptians wanted to write fractions with a numerator greater than 1, they wrote them as the sum of different unit fractions.

For example, they wrote $\frac{3}{4}$ as $\frac{1}{2} + \frac{1}{4}$ and $\frac{33}{40}$ as $\frac{1}{2} + \frac{1}{5} + \frac{1}{8}$

The unit fractions had to be different, so $\frac{2}{3}$ could not be written as $\frac{1}{3} + \frac{1}{3}$

What different unit fractions total to make $\frac{2}{3}$? Can you find more than one way of doing this?

What about other fractions with a numerator of 2? In each case, try to use as few unit fractions as possible.

Extend your investigation to look at numerators of 3, 4 or 5 and beyond.

3.6 Decimals

3.6.1 Rounding decimals

Rounding a value produces a simpler number. The rounded number is not quite as accurate as the original value but is close enough to be useful.

Explore 3.11

Here are some decimals shown on a number line with the accuracy and number they round to.

7.2 rounds to 7 to the nearest whole number.

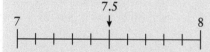

7.5 rounds to 8 to the nearest whole number.

Research why mathematicians round 5 up even though it is exactly half-way.

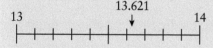

13.621 rounds to 14 to the nearest whole number.

1.308 rounds to 1.31 to the nearest 2 decimal places (2 d.p.).

2.996 rounds to 3.00 to the nearest 2 d.p.

Can you explain how to round to:

- the nearest whole number
- one decimal place
- two decimal places?

Population and attendance figures are usually reported in millions or thousands. Why do you think that is?

Worked example 3.8

a Round 23.5184 to the nearest whole number.

b Round 23.5184 to the nearest 1 decimal place.

c Round 23.5184 to the nearest hundredth.

d Round 4.666 centimetres to the nearest millimetre.

Solution

a 23.5184 when rounded to the nearest whole number has a cut-off point after the ones column:

23|.5184

The digit in the column to the right of the cut-off point is 5.

Round up if the next digit is 5 or more.

23.5184 rounds to 24 to the nearest whole number.

b 23.5184 when rounded to the nearest 1 decimal place has a cut-off point after the tenths column:

23.5|184

The digit in the column to the right of the cut-off point is 1.

Round down if the next digit is less than 5.

23.5184 rounds to 23.5 to the nearest 1 decimal place.

c 23.5184 when rounded to the nearest hundredth has a cut-off point after the hundredths column:

23.51|84

The digit in the column to the right of the cut-off point is 8.

Round up if the next digit is 5 or more.

23.5184 rounds to 23.52 to the nearest hundredth.

d 4.666 centimetres = 46.66 millimetres   10 mm = 1 cm

Rounding 46.66 mm to the nearest whole number gives 47 mm.

 Reflect

Apply what you have discovered so far to summarise how to round to any given number of decimal places.

Practice questions 3.6.1

1   Round these numbers to the nearest whole number.

    a   8.7   b   4.4   c   25.1   d   16.62

    e   45.95   f   0.6712   g   2.908   h   245.099

2   Round these numbers to one decimal place.

    a   5.42   b   8.671   c   36.61   d   2.4991

    e   0.091   f   0.05   g   10.049   h   458.129

3   Round these numbers to the given degree of accuracy.

    a   254.2698 to 1 decimal place

    b   0.5871 to 2 decimal places

    c   2.1994 to the nearest tenth

    d   15.1515 to the nearest hundredth

    e   9.99 to 1 decimal place

    f   8.809 to the nearest tenth

    g   0.5076172 to the nearest hundredth

    h   3.5264 to the nearest hundredth.

4   Each calculator display gives numbers as currency, in dollars and cents.
    Round each answer to the nearest cent.

    a   23.25487   b   4587.735489

    c   129.4995217   d   23564821.3

    e   90.548723569810   f   2300.715887

5   Round each of the calculator displays in question 4 to the nearest
    dollar.

6   Round each value to the level of accuracy given.

    a   1.2763 metres to the nearest centimetre

    b   2.8457 litres to the nearest millilitre

     c   47.45 centimetres to the nearest millimetre

     d   5.08409 kilometres to the nearest metre

     e   56.112548 kilograms to the nearest gram.

7   Which of these numbers are the same when rounded to 1 decimal place?

     13.99   13.94   13.9   13.95   13.908   13.09

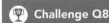 **Challenge Q8**

8   Write down a number that rounds to 15 when rounded to 1 decimal place and to the nearest whole number.

 **Challenge Q9**

9   Nahla says, 'I'm thinking of a number. When I round my number to 1 decimal place, I get 7.8.'

     a   If Nahla's number was rounded up, write down one value that it could have been.

     b   If Nahla's number was rounded down, write down one value that it could have been.

 **Challenge Q10**

10   Write four numbers that round to 3.45.

 **Challenge Q11**

11   Copy and complete the number 23.25 ☐☐☐ so that it rounds to 23.26 to 2 decimal places and 23.260 to 3 decimal places.

 **Challenge Q12**

12   Muna, Noor and Samira inherit \$10 000 from their grandfather.

     a   How much does each person receive?

     b   Explain why it is not possible for each to receive an equal amount.

 **Challenge Q13**

13   A number is rounded to 1 decimal place to get 19.5.

     a   What is the greatest value it could be?

     b   What is the smallest value it could be?

3.6.2 The four operations with decimals

Explore 3.12

Leila, Ola and Sophie work out 0.4×0.2

Leila writes:	Ola writes:	Sophie writes:
$0.4 \times 0.2 = 0.8$	$0.4 \times 0.2 = 0.08$	$0.4 \times 0.2 = 0.008$

Check who is correct. Can you explain why they are correct?

Work out $0.015 \div 0.03$. Can you explain your method?

Does your method work for $0.172 \div 0.04$?

Worked example 3.9

Work out:

a   15.73 + 2.5   **b**   57.3 × 2.4

Solution

a   Write 15.73 + 2.5 in columns:

Tens	Ones	·	$\frac{1}{10}$	$\frac{1}{100}$
1	5	·	7	3
	2	·	5	0
1	8	·	2	3

+

$_1$

You can use zeros as place holders.

15.73 + 2.5 = 18.23

b   To work out 57.3 × 2.4, first work out 573 × 24

One method is to use long multiplication.

57.3 = 573 ÷ 10 and 2.4 = 24 ÷ 10

Therefore,

57.3 × 2.4 = 573 × 24 ÷ 10 ÷ 10

            = 13 752 ÷ 10 ÷ 10

            = 137.52

```
            5    7    3
                 2    4
        2   ₂2   ₁9   2
   +  1  ₁1    4    6    0
   ─────────────────────
      1   3    7    5    2
                 ₁
```

Reflect

An alternative method for multiplying decimals is the method of Gelosia. This method is believed to have originated in India, and it is known by many other names today.

This is the layout for solving 57.3 × 2.4 using the Gelosia method:

Try to explain how and why the Gelosia method works.

What alternative methods can you find to work out 57.3 × 2.4?

Which method do you prefer?

Practice questions 3.6.2

1   Work out:

   a   0.54 + 5.07   b   7.045 − 2.96   c   254.05 + 4.9   d   0.5 × 0.4

   e   1.2 × 0.7   f   2.3 × 3   g   100 − 2.891   h   4.8 ÷ 0.4

   i   2.4 ÷ 0.12   j   0.75 ÷ 1.5   k 0.7^2 l   560 ÷ 0.08

2   Work out:

   a   (0.3 + 1.5) × 0.3   b $0.4^2 − 0.05$

   c   12.8 + 4.6 × 0.2   d   (15.5 + 12.7) ÷ (4.5 − 4.2)

   e $(0.205 − 0.095)^2$ f $0.8^2 ÷ 0.04$

   g   (25.091 − 4.591) × (0.42 ÷ 0.7)   h $(0.3^2 − 0.2^2)^2$

3   Do 5.8 × 2.7 and 0.58 × 270 give the same answer?

   Explain your mathematical reasoning.

4   Use the fact that 4.5 × 2.3 = 10.35 to decide if each calculation below is true or false.

   a   45 × 0.23 = 10.35   b   4.5 × 23 = 1.035

   c   4.5 × 23 = 103.5   d   2.3 × 450 = 1035

   e   230 × 0.45 = 103.5   f   0.45 × 0.23 = 1.035

5   Use the fact that 0.7 × 3.05 = 2.135 to copy and complete these calculations:

   a   7 × ☐ = 2.135   b   0.07 × ☐ = 21.35

   c   0.7 × ☐ = 2135   d   ☐ × 30.5 = 2.135

   e   ☐ × 0.305 = 21.35   f   ☐ × 305 = 0.2135

6   Use the fact that 5.6 × 0.18 = 1.008 to write down three other calculations and their answers.

7   Ali and Waleed are racing snails.

   Ali's snail took 3.7 minutes to complete the course.

   Waleed's snail took 4.08 minutes to complete the course.

   How much quicker was Ali's snail?

8   Almira has four lengths of silk.
    Each length is 1.6 m long.
    Almira needs 6.5 m of silk.
    Does she have enough silk?
    Explain your answer.

9   A group of six people go to a restaurant.
    The bill at the end of the evening is $91.50.
    They share the bill equally.
    How much does each person pay?

10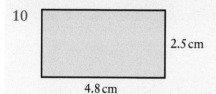

    a   What is the perimeter of this rectangle?

    b   What is the area of the rectangle?

11 The rectangle in the diagram has an area of 15.8 cm².

    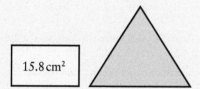

    The triangle has an area four and a half times bigger than the
    rectangle.
    Work out the area of the triangle.

12 Alia, Leila and Maya all compete in a long jump event.
    Alia jumps 448.7 cm.

    Alia jumps $3\frac{1}{2}$ times the length of Leila.

    Maya jumps 102.7 cm further than Leila.
    How far does Maya jump?

13 In this pyramid, each block in the upper layers
    can be found by adding together the two
    numbers below it.

    Copy and complete the number pyramid.

    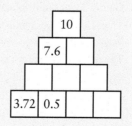

14 Copy and complete the multiplication grid.

×		0.3		1.2
		0.21		
5.7			17.1	
	1.16			6.96

 Connections

Game: Total 2.0

This is a two-player game.
You need a copy of the decimals and the grid:

Player 1
0.2
0.4
0.6
0.8
1.0

Player 2
0.2
0.4
0.6
0.8
1.0

Players take turns to write one of their decimals in the grid. Once you have used a decimal, cross it off your list. You cannot use a decimal more than once.

The winner is the first player to finish a row, column or diagonal that sums to 2.0.

 Investigation 3.7

Find two decimals with a sum of 1. What is the product of your two decimals?

Repeat for a few pairs of decimals.

What are the greatest and least products you can get?

Thinking skills

 Connections

Game: Target multiplication

This is a two-player game, and you will need one calculator.
Begin by agreeing on a target number. Let us say you have chosen a target number of 50.

The first player enters a number less than the target number into the calculator.

The second player multiplies by any number to get as close to the target number as possible.

This process is repeated.

The winner is the first player to get within 0.5 of the target number.

I can identify and use number properties for addition and multiplication.

I can list the factors and multiples of numbers.

I can find the HCF and LCM of two numbers.

I can identify square and cube numbers.

I can work out a number to a given power.

I know and can use the laws of indices for multiplication and division.

I can identify and use prime numbers and composite numbers.

I can write a number as a product of its prime factors.

I can find equivalent fractions.

I can simplify fractions.

I can order fractions.

I can change an improper fraction to a mixed number and vice versa.

I can find a fraction of a quantity.

I can add, subtract, multiply and divide fractions and mixed numbers.

I can round decimals to the nearest whole number or to a given number of decimal places.

I can add, subtract, multiply and divide decimals.

❓ Check your knowledge questions

1 Work out the values of:

a $32\,876 \times 1$
b $567\,418 \times 0$
c $7 \times 5 \times 23$

2 Write down whether each equation is true or false.

a $134 + 62 = 62 + 134$
b $57 - 16 = 16 - 57$
c $32 \times 497 = 497 \times 32$
d $72 \div 8 = 8 \div 72$

3 Write down the missing number that will make each equation true.

a $\boxed{} \times 925 = 925$
b $23\,584 \times \boxed{} = 0$
c $\boxed{} + 623 = 623 + 59$
d $34 \times \boxed{} = 27 \times 34$

4 Write down the missing number that will make each equation true.

a $3 \times 7 \times 40 = \boxed{} \times 3$
b $126 + 1 = \boxed{} + 1 + 53$

5 Without using a calculator, use number properties to find the answers quickly to these calculations:

a $2 \times 35 \times 5$
b $159 \times 46 \times 0$
c $67 + 254 + 63$

6 Write down the first six multiples of:

a   3
b   6
c   15

7 List all the factors of:

  a 10 b 18 c 49 d 80

8 Write down whether each statement is true or false.

  a 3 is a factor of 425 b 132 is a multiple of 12

  c 6 is a factor of 333 d 4 is a factor of 300.

9 Work out the highest common factor of 104 and 156.

10 Work out the lowest common multiple of 15 and 18.

11 Look at this set of numbers.

2, 4, 6, 8, 10, 11, 12, 15, 16, 20, 36
List the:

  a factors of 16 b multiples of 3

  c square numbers d prime numbers.

12 Write each of these as a base number raised to a power.

  a $4 \times 4 \times 4$ b $9 \times 9 \times 9 \times 9 \times 9$

13 Write each of these in expanded form.

  a 7^4 b 2^7

14 Work out:

  a 6^2 b 12^2 c 13^2

15 Calculate:

  a 4^3 b 7^3 c 12^3

16 Work out:

  a 3^6 b 2^7 c 15^4

17 Write down the values of:

  a 6×10^2 b 8×10^5 c $5^2 \times 10^4$ d $7^2 \times 10^3$

18 Write down the values of:

  a $3^2 + 5^2$ b $4^2 \times 3^3$

  c $(20 \div 4)^3$ d $(4^3 - 7^2)^2 - 10^2$

19 Write down the missing numbers.

a $\boxed{}^2 = 25$

b $\boxed{}^2 = 144$

c $\boxed{}^3 = 64$

d $\boxed{}^3 = 1000$

20 Write down two square numbers that total 100.

21 Write each of these as a single power.

a $7^4 \times 7^3$

b $11^7 \times 11^8$

c $10^3 \times 10^{-1}$

d $10^3 \times 10^{-5}$

e $3^8 \div 3^5$

f $10^9 \div 10^{-7}$

g $5^4 \div 5^{-3}$

h $\dfrac{13^9}{13^4}$

i $\dfrac{6^{10}}{6^7}$

22 Write each number as a product of its prime factors.

a 36

b 100

c 150

23 Write down the fraction that is *not* equivalent to the other fractions.

$\dfrac{8}{12}, \dfrac{12}{18}, \dfrac{20}{30}, \dfrac{24}{35}, \dfrac{30}{45}$

24 Simplify these fractions.

a $\dfrac{24}{32}$

b $\dfrac{25}{30}$

c $\dfrac{85}{100}$

d $\dfrac{48}{60}$

25 Place one of these signs, <, = or >, between each fraction pair to make the statement true.

a $\dfrac{7}{10}$ $\boxed{}$ $\dfrac{17}{25}$

b $\dfrac{3}{8}$ $\boxed{}$ $\dfrac{7}{20}$

26 Place these fractions in ascending order.

$\dfrac{7}{10}, \dfrac{3}{5}, \dfrac{5}{8}, \dfrac{13}{20}$

27 Write $\dfrac{19}{6}$ as a mixed number.

28 Write $3\dfrac{3}{4}$ as an improper fraction.

29 Work out the fractional amount of each number.

a $\dfrac{2}{5}$ of 60

b $\dfrac{7}{8}$ of 56

c $\dfrac{3}{10}$ of 1000.

30 Work out each of these, giving each of your answers in smaller units if necessary.

a $\dfrac{1}{10}$ of \$2

b $\dfrac{3}{4}$ of 3.6 litres

c $\dfrac{4}{5}$ of 1 hour.

31 $\frac{3}{5}$ of a number is 27.

What is the number?

32 Work out each calculation.

Give each answer in its simplest form.

a $\frac{4}{5} + \frac{7}{9}$ b $\frac{7}{8} - \frac{2}{3}$ c $\frac{6}{7} \times \frac{14}{15}$ d $\frac{5}{8} \div \frac{3}{4}$

e $3\frac{1}{2} + 2\frac{3}{5}$ f $5\frac{2}{3} - 1\frac{7}{8}$ g $1\frac{7}{10} \times 3\frac{1}{3}$ h $3\frac{9}{10} \div 1\frac{1}{2}$

33 Copy and complete this calculation.

$$2\frac{3}{4} - \boxed{}\frac{\boxed{}}{\boxed{}} = \frac{4}{5}$$

34 Round these numbers to the given degree of accuracy.

a   15.32 to the nearest whole number

b   7.95 to 1 decimal place

c   0.364 to 2 decimal places

d   2.3456 to the nearest tenth.

35 Round 22.654 centimetres to the nearest millimetre.

36 Write four numbers that round to 2.25, to 2 decimal places.

37 Work out these calculations.

a   156.08 + 3.7   b   1.8 × 0.7   c   10 − 7.541   d   0.56 ÷ 0.8

38 Work out these calculations.

a $0.6^2 - 0.07$ b   5.8 + 6.4 × 0.3   c $(5 - 0.1^2)^2$

39 Use 0.4 × 7.03 = 2.812 to copy and complete these calculations.

a   0.04 × $\boxed{}$ = 28.12

b   0.4 × $\boxed{}$ = 2812

c $\boxed{}$ × 703 = 0.2812

Geometry

4

4 Geometry

KEY CONCEPT

Relationships

RELATED CONCEPTS

Equivalence, Space, Systems

GLOBAL CONTEXT

Orientation in space and time

Statement of inquiry

Using a system of measuring to determine an object's position in space can help us to understand its relationships with other objects.

Factual:

- What is the sum of the angles at a point and on a straight line?
- What are the names and definitions of the different angle types?
- What are the names and definitions of the different triangle types?

Conceptual:

- What are the similarities and differences between different types of triangles?
- Why do we have naming conventions for angles and lines?

Debateable:

- Why can classifying objects by their properties and attributes be useful?

Do you recall?

1 Which of these diagrams shows a $\frac{1}{4}$ turn?

A B C D

2 State the size of the turn in the other diagrams in question 1.

3 Which of these angles is a right angle?

A  B C

4 Which diagram shows a pair of perpendicular lines?

A  B C

5 How many degrees are there in a right angle?

6 How many degrees are there in a full turn? How many in a half turn?

7 What are the four compass directions?

4.1 Review and practice – measuring and drawing angles

4.1.1 Measuring angles

Explore 4.1

What is the same and what is different about these two angles?

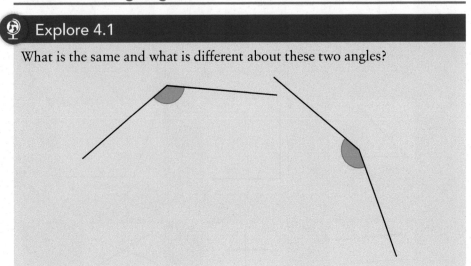

An **angle** measures the size of a turn.
Angles are measured in degrees.

An angle is between two lines. The point where the
two lines meet is the **vertex** of the angle.

You can use a protractor to measure an angle.
There are two scales on the protractor.

Place the protractor on the angle, with the base line of the semicircle along
one line of the angle.

vertex

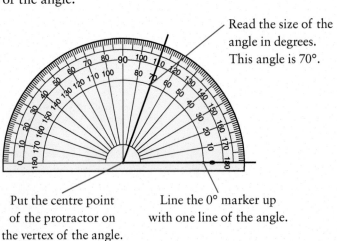

Read the size of the
angle in degrees.
This angle is 70°.

Put the centre point
of the protractor on
the vertex of the angle.

Line the 0° marker up
with one line of the angle.

Angles are classified by their size.

Acute angle	Right angle	Obtuse angle
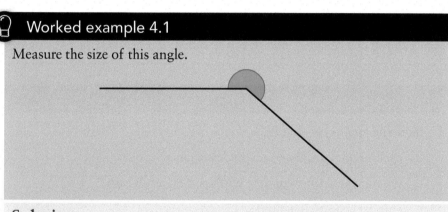		
Less than 90°	90°	Between 90° and 180°

Reflex angle	Straight angle	Revolution
Between 180° and 360°	180°	360°

⑦ Reminder

- A revolution is one complete turn.
- A straight angle is a half turn.
- A right angle is a quarter turn.
- The arms of a straight angle form a straight line.
- The special symbol for a right angle is
- The right angle, straight angle and angle of revolution have sizes: 90°, 180° and 360° respectively.

◎ Fact

The ancient Babylonians, Greeks and Indians all divided circles into 360°, though we do not know their reasoning. One suggestion is that they thought it took 360 days for the Sun to orbit the Earth. Another is that they chose 360 because it has many factors, so can be divided easily into equal angles.

♀ Worked example 4.1

Measure the size of this angle.

Solution

Method 1

You can use a 360° protractor.

Place the zero line of the protractor on one line of the angle.

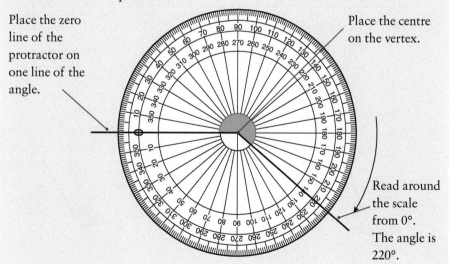

Place the centre on the vertex.

Read around the scale from 0°. The angle is 220°.

Method 2

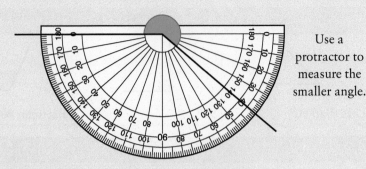

Use a protractor to measure the smaller angle.

The smaller angle is 140°.

Subtract the angle from a full turn: 360° − 140° = 220°

Reflect

Look back at Explore 4.1.

How can you describe the similarities and differences between the two angles?

Practice questions 4.1.1

1   State the angle type that is:

   a   greater than a right angle but less than a straight line

   b   greater than a straight line but less than a full turn

   c   smaller than a right angle.

2
| 49° | | 249° | | 127° | | 27° | |
| | 72° | | 294° | | 94° | | 217° |

Write down the angles from this box that are:

   a   acute angles   b   obtuse angles   c   reflex angles.

3   a   State whether each angle is obtuse or acute.

   b   Measure the size of each angle.

   i    ii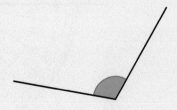

iii 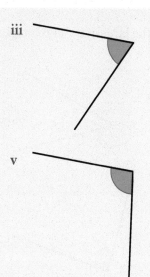   iv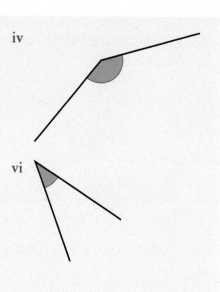

v   vi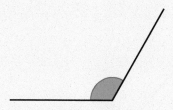

4   Smita says this angle measures 60°.

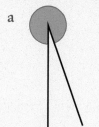

a   Explain the mistake she has made.

b   Find the size of the angle.

5   Measure the size of each angle.

a    b

c 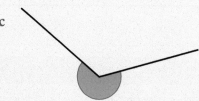   d

6   What is the difference in size between the largest and the smallest of these angles?

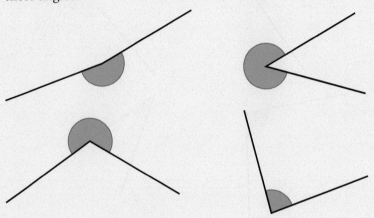

7   Measure the angles marked with letters in each shape.

a

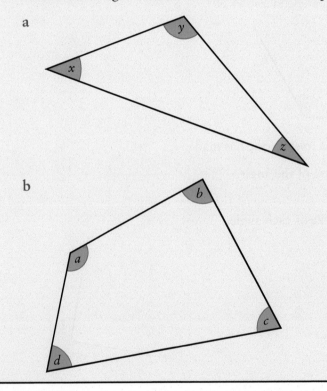

b

4.1.2 Drawing angles

Explore 4.2

You know how to measure angles with a protractor. How could you use a protractor to draw angles accurately?

Explore how to draw acute, obtuse and reflex angles.

The four cardinal directions (the main directions) of the compass are:

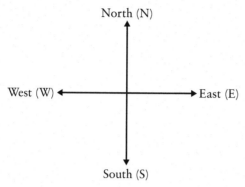

The angle between North and East is a right angle, 90°.

In diagrams, you draw North straight up the page.

You can give more accurate directions using eight points of the compass:

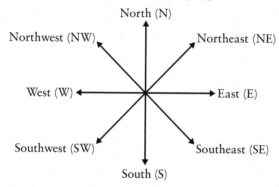

The angle between North and Northwest is half of 90° = 45°

Worked example 4.2

The diagram shows the position of two villages.

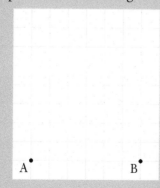

A road runs North from village A.

Another road runs Northwest from village B.

Mark the point P where the roads cross.

Solution

Use a ruler to draw a straight line segment from A to B.

Draw a North line from A, straight up the page.

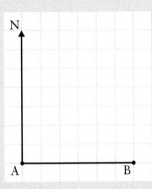

Draw a North line at B.

Measure 45° from the North line, to give the Northwest direction.

Draw a line Northwest from B to cross the line from A.

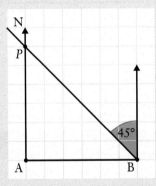

Label the point P where the lines cross.

⇄ Reflect

What are the names of the cardinal directions in your first language?
In English, the phrase Never Eat Sour Watermelon can help you remember the cardinal directions and their positions. You could make up your own phrase.

A sextant is used to measure the angle between a star and the horizon. It is highly accurate, which makes it useful for navigation.

1 Copy this diagram.

Label the angle between each
line and the one next to it.

Write in the names of the
eight compass points.

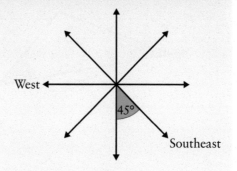

2 a State whether each angle below is obtuse or acute.

b Draw each angle.

i 40°   ii 130°   iii 55°   iv 125°   v 117°   vi 62°

3 Raoul is trying to draw an angle of 73°.

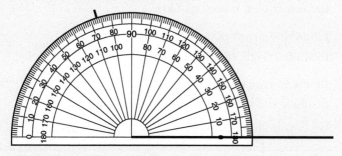

Explain the mistake he has made.

4 Draw an angle of 50°.
Label the angle of 360° − 50° = 310° on your diagram.

5 Draw each angle.

a 320°   b 210°   c 275°   d 193°

6 The diagram shows the position of
two villages.

A road runs Southeast from village A.

Another road runs South from village B.

Copy the diagram and mark the point *P*
where the roads cross.

A•   B•

Hint Q7a

Draw the 7 cm line first.

7 Draw each triangle accurately. Use a ruler and protractor.

a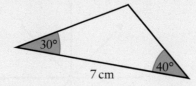

b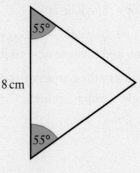

🏆 **Challenge Q8**

🎓 **Thinking skills**

8 Sophie leaves her home and walks East for 4 km.

Then she turns and walks Southwest until she is directly South of her home.

a Draw an accurate diagram on squared paper to represent Sophie's walk.

Let one square represent 1 km.

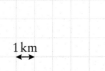

b Find the shortest possible distance for Sophie to walk straight home.

c By measuring the distances on your diagram, estimate the total distance she walked.

Give your answers in km.

4.2 Working with angles

4.2.1 Naming angles

vertex

❓ **Reminder**

An angle is between two lines. The point where the two lines meet is called the vertex.

When the lines of an angle are labelled with letters:

• you can name each side using the letters at its ends

• you can name the angle by using the letter at the vertex.

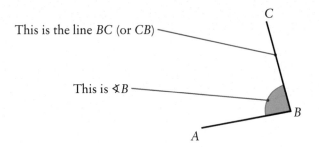

This is the line *BC* (or *CB*)

This is ∢*B*

In this diagram, there are two angles with vertex *B*.

You can name the angles using three letters.

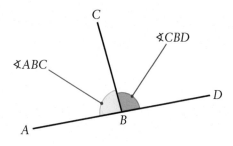

∢*ABC*

∢*CBD*

For three-letter labels, the vertex is the middle letter.

∢*ABC* = 85° tells you that when you go from *A* to *B* to *C*, you turn through an angle of 85° at *B*.

In a diagram, equal numbers of arcs show equal angles.

In this trapezium:

∢*A* = ∢*D*

∢*B* = ∢*C*

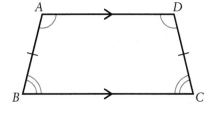

You can also label angles with lower case letters:

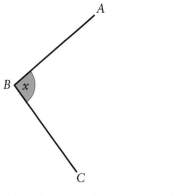

In a diagram, letters labelled with the same letter are equal.

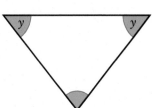

Practice questions 4.2.1

1 Name the angle marked with an arc in each diagram.

a
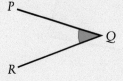

b
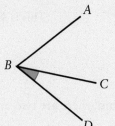

c
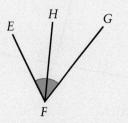

d
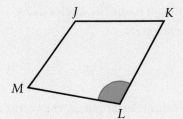

e
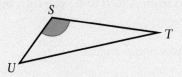

f
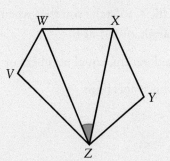

2 Write the three-letter name of the right angle shown in each diagram.

a
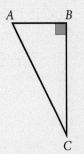

b
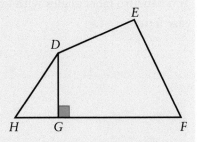

c
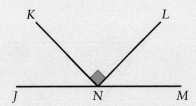

d
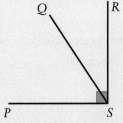

3   Max, Saleem and Kobe name this angle.

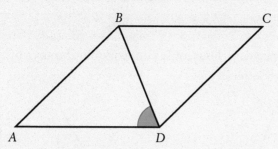

Max: ⊰*BDA*   Saleem: ⊰*ADB*   Kobe: ⊰*BAD*

Who is correct? Explain how you know.

4   Draw accurate diagrams to show these angles.

a   ⊰*FXQ* = 20°   b   ⊰*ARG* = 115°   c   ⊰*DNP* = 335°

5   Name each type of angle in this **quadrilateral**.

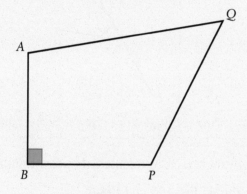

a   ⊰*PBA*   b   ⊰*BPQ*   c   ⊰*AQP*

6   Measure these angles in quadrilateral *ABCD*.

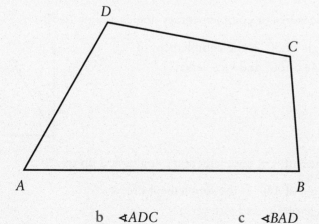

a   ⊰*B*   b   ⊰*ADC*   c   ⊰*BAD*

> **Reminder**
>
> A quadrilateral is a four-sided shape.

4.2.2 Calculating angle sizes

 Explore 4.3

Without measuring, how could you work out the sizes of the angles labelled with letters?

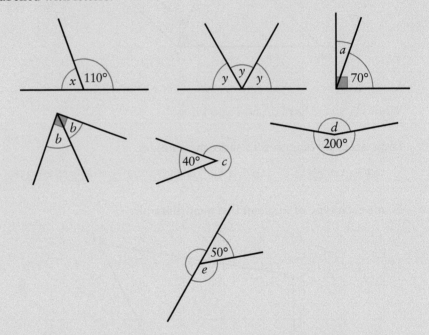

Why might measuring the angles not give the correct answers?

You can use these angle facts to calculate the size of unknown angles.

- Angles on a straight line add up to 180°.

- Angles in a right angle add up to 90°.

- Angles at a point add up to 360°.

- The measures of complementary angles add up to 90°.

  An angle of 30° is the complement of an angle of 60° and vice versa.

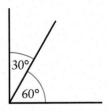

- The measures of supplementary angles add up to 180°.

  An angle of 130° is the supplement of an angle of 50° and vice versa.

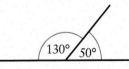

a Find the size of ∢*ABC*.

b Work out the size of ∢*FGH*.

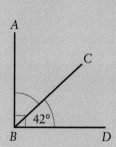

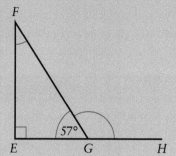

Give reasons for your answers.

Solution

Understand the problem

We need to calculate the sizes of the angles, and give the reasons or facts we use in the calculations.

Make a plan

First we will identify the angle we need to find. Then we will use an angle fact to calculate the angle.

Carry out the plan

a ∢*ABC* = 90° − 42° = 48°   (∢*ABC* and ∢*CBD* are complementary angles)

Another way of explaining this reasoning is: angles in a right angle add up to 90°.

b ∢*FGH* = 180° − 57° = 123°   (∢*EGF* and ∢*FGH* are supplementary angles)

Another way of explaining this reasoning is: angles on a straight line add up to 180°.

Look back

Do the answers make sense?

a The angle to find is an acute angle in the diagram, and the answer 48° is an acute angle.

b The angle to find is an obtuse angle in the diagram, and the answer 123° is an obtuse angle.

Reflect

Look back at the angles you found in Explore 4.3.

How would you give your reasons for each one?

Why do you think it is useful to give your reasons?

 Communication skills

 Self-management skills

 Investigation

Which pair of lines are parallel?

Which pairs of lines are perpendicular?

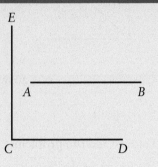

Draw a pair of parallel lines on squared paper.

Now draw a straight line that crosses both parallel lines.

A straight line crossing two parallel lines is called a **transversal**.

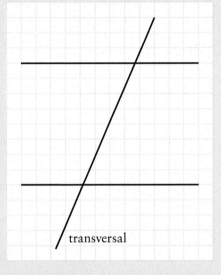

transversal

Measure and label all the angles on your diagram where the lines intersect.

Use colour to show equal angles on the diagram.

Are any angles complementary? Supplementary?

Repeat for at least two more pairs of parallel lines and a transversal.

What do you notice about the angles in parallel lines?

What if:

• the parallel lines are vertical

• the transversal is perpendicular to the parallel lines?

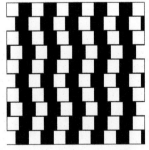

Do you think the lines that make the tops and bottoms of these squares are parallel? Look carefully!

1   Work out the size of each angle labelled with a letter.

a

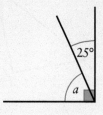

25°

a

b

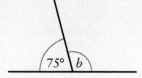

75°   b

c

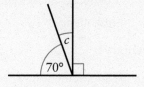

c

70°

d

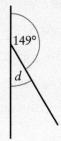

149°

d

e

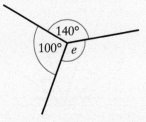

140°

100°   e

f

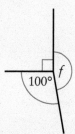

100°   f

g

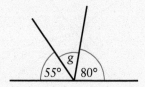

g

55°   80°

h

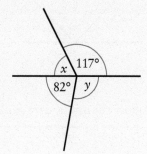

x   117°

82°   y

2   a   Work out the size of ∢BCD.

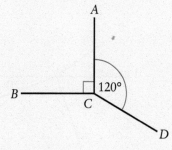

A

120°

B

C

D

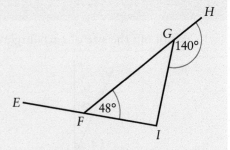

b Work out the size of:

    i ∢*EFG*

    ii ∢*FGI*

    Give reasons for your answers.

3 a Can two acute angles be:

    i supplementary   ii complementary?

    b Can two obtuse angles be complementary?

    c Can a reflex angle and an acute angle be supplementary?

    Explain your reasoning.

4 Find the sizes of the angles labelled with letters. Give reasons for your answers.

a

b

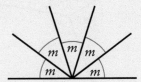

c

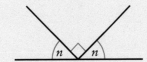

d

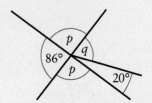

5 a Sort the angle sizes in this box into pairs of supplementary angles.

70°	95°	36°	165°	85°	105°
15°	144°	110°	20°	75°	154°

    b Write down sizes of supplementary angles for any angles left over.

Hint Q6

In a 'show that' question, show working and give reasons for each step.

6 Show that ∢*FCA* = 2∢*ACB*

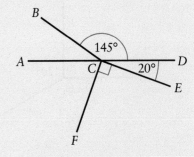

7   Make a set of 20 'supplementary angle dominoes'.

Play a dominoes game with a classmate.

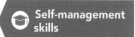

8   Quadrilateral *PQRS* has vertices labelled in alphabetical order around the shape.

Use these clues to help you make an accurate drawing of *PQRS*.

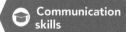

- ∡*P* is a right angle

- ∡*RSP* = ∡*P*

- ∡*PQR* = 70°

- ∡*PQR* and ∡*QRS* are supplementary.

4.3   Triangles

4.3.1 Classifying triangles

A triangle is a 2D shape (or **plane** shape) with three sides. The sides are straight lines. A triangle has three angles and three **vertices**.

A **vertex** of a 2D shape is where two sides meet. The plural of vertex is vertices.

You can classify triangles by their side and angle properties.

Equilateral	Isosceles	Scalene
Three equal sides Three equal angles	Two equal sides Two equal angles, at the bases of the two equal sides	No equal sides No equal angles

> **Fact**
>
> The name 'isosceles' comes from two Greek words, meaning 'equal' and 'leg'.

You can also describe or classify triangles by their angles:

Acute-angled	Right-angled	Obtuse-angled
All three angles are acute.	One angle is a right angle.	One angle is obtuse.

Explore 4.4

Draw diagrams to explain your answers to these questions.

Can a triangle:

- have two right angles

- have two obtuse angles

- be scalene and acute-angled

- be right-angled and scalene

- be isosceles and right-angled

- be isosceles and obtuse-angled

- be equilateral and obtuse-angled?

 Practice questions 4.3.1

1   Classify each triangle. Give all the possible classifications for each one.

a    b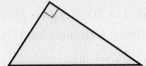

c 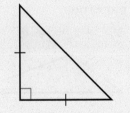   d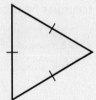

e    f

2   Measure the sides and angles in each triangle.

Hence, classify each triangle.

a    b

c 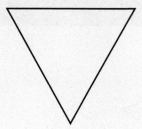   d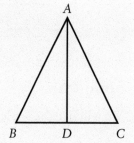

3   Here is triangle ABC.

    a   Which sides of triangle ABC are equal?

    b   Which angles of triangle ABC are equal?

    c   What type of triangle is ABC?

    d   What type of triangle is ADC?

    e   Is AD a line of symmetry of triangle ABC?
       Explain how you know.

Reminder

If you fold a shape along a line of symmetry, both halves fit exactly on to each other.

4   Use a pair of compasses to draw a circle. Mark the centre and any two points on the circle. Join these three points with straight lines.

    What type of triangle have you drawn?

    Repeat with another two points on the circle, and the centre.
    What do you notice? Can you explain why this happens?

4.3.2 Angles in a triangle

 Explore 4.5

Draw any triangle and cut it out.

Label the angles 1, 2 and 3.

Tear the vertices off your triangle.

Fit the angles together side by side, like this:

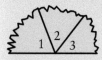

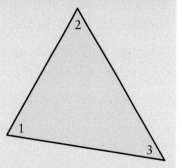

What do you notice? What does this tell you about the angles in your triangle?

Repeat with at least two more triangles to check your results.

The angle sizes in a triangle add to 180°.

Another way of saying this is: the angle sum of a triangle is 180°.

Self-management skills

Communication skills

Fact

Geometry means 'measurement of the Earth'. The ancient Egyptians used their understanding of triangles to build the pyramids with their four isosceles triangular faces. Thales of Miletus (624–547 BCE) is believed to be the first Greek to study geometry, based on Egyptian and Babylonian knowledge. Later, Euclid of Alexandria (325–265 BCE) wrote *The Elements* – 13 books on geometry which have been used in schools ever since, and are still in print.

Worked example 4.4

Find the sizes of the angles in this triangle.

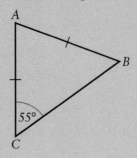

Solution

We will use the angle facts we know to identify any equal sides or angles.

$\sphericalangle ABC = \sphericalangle ACB = 55°$ (base angles in an isosceles triangle)

$180° - 55° - 55° = 70°$ (angle sum in a triangle = 180°)

$\sphericalangle CAB = 70°$

Reflect

Why is it important to state your reasons when finding the angle sizes in a triangle?

How do you know which angle in triangle *ABC* is equal to the 55° angle?

Why can't you just measure the angles in the diagram?

When you look up at an object, the angle your line of sight makes with the horizontal is called the **angle of elevation**.

When you look down at an object, the angle your line of sight makes with the horizontal is called the **angle of depression**.

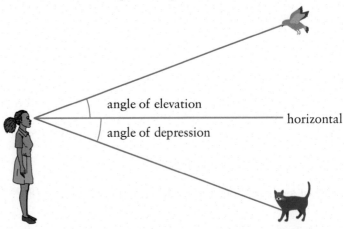

Worked example 4.5

The diagram shows the position *A* of a person on a cliff, and the position *B* of a surfer in the sea.

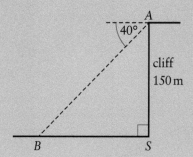

The angle of depression from *A* to *B* is 40°.

Find the angle of elevation from *B* to *A*.

Solution

∢ *BAS* = 50° (complementary angles)

∢ *ABS* = 40° (sum of angles in a triangle is 180°)

> **Hint**
>
> The cliff is vertical, so it is at 90° to the horizontal line at *A*.

Practice questions 4.3.2

1   Find the size of the angle labelled with a letter in each triangle.

a
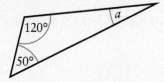

b
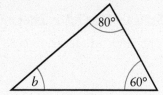

c
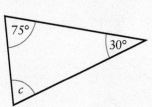

d
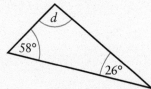

e
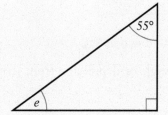

f
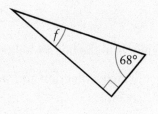

2 Work out the size of the angles in an equilateral triangle.

3 Find the sizes of the angles labelled with letters in these triangles.

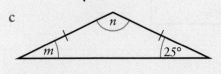

a

63°   g

h

b

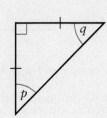

j

40°

k

c

n

m   25°

d

q

p

4 The diagram shows a ladder leaning against a wall.

Find the angle between the ladder and the wall.

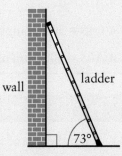

wall   ladder

73°

5 Saj is looking up at a hot air balloon, which is directly above his house.

The angle of elevation is 36°.

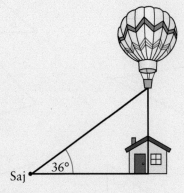

Saj   36°

Jenny is in the hot air balloon looking down at Saj.

a Sketch a diagram and show the angle of depression from Jenny to Saj.

b Work out the size of the angle of depression.

6   Find the sizes of the angles labelled with letters in these diagrams.
    Give reasons.

    a

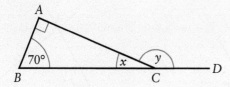

    b

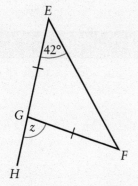

    c

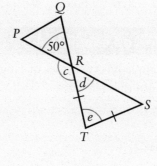

7   One angle in an isosceles triangle is 66°. What could the other two
    angles be? Find two possible answers.

🏆 Challenge Q7

8   Calculate the angle between one side of a square and the diagonal of
    the square. Show your reasoning clearly.

🏆 Challenge Q8

4.3.3 Drawing triangles accurately

🌐 Explore 4.6

Draw three different triangles. Measure and label the side lengths and angles.

In each triangle, what do you notice about the angle opposite

- the longest side

- the shortest side?

Which is the largest angle in a right-angled triangle? Which is the longest
side in a right-angled triangle?

You can draw a triangle accurately using a
ruler and protractor, if you know some
side lengths and angles.

The following steps can be used to draw
triangle *ABC*.

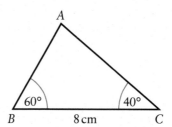

Use your ruler to draw the base, 8 cm long.

Draw the 60° angle at B.

Draw the 40° angle at C. Extend the lines from B and C so they intersect at A.

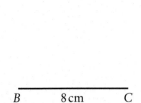

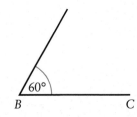

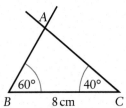

Worked example 4.6

A triangle has two sides of lengths 4 cm and 5 cm, with an angle of 60° between them.
Draw this triangle accurately.

Solution

First, sketch the triangle.

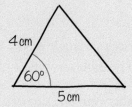

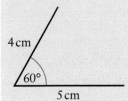

Start by drawing one of the sides of known length.

Draw the 60° angle.

Draw the line exactly 4 cm long.

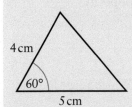

Join the ends of the two lines.

Reflect

Why is it helpful to sketch the triangle first?
Does it matter which side of the triangle you draw first?
Why is it important to use a sharp pencil?

1   Draw these triangles accurately.

a

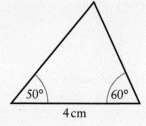

50°   60°
4 cm

b

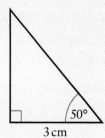

50°
3 cm

c

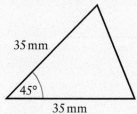

35 mm

45°
35 mm

d

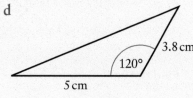

3.8 cm
120°
5 cm

2   Draw this triangle accurately. Measure sides *LN* and *MN*.

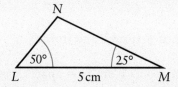

N

50°   25°
L   5 cm   M

3   Make an accurate drawing of an equilateral triangle with sides of 6 cm.

4   Make an accurate drawing of an isosceles triangle with base 5 cm and base angles 55°.

5   Draw accurately a right-angled triangle with perpendicular sides 3 cm and 4 cm. Measure the length of the third side.

6   Maisie stands 12 m from a tree. She measures the angle of elevation to the top of the tree to be 65°.

Draw an accurate diagram, with 1 cm on the diagram representing 1 m in real life.

Use your diagram to find the height of the tree.

65°
12 m

🏆 Challenge Q7

7   How many different triangles can you construct with a 4 cm side, a 5 cm side and a 45° angle?

8   Two boats, *A* and *B*, leave a harbour and sail in different directions.

Boat *A* sails Northwest for 5 km.

Boat *B* sails East for 7 km.

Copy this diagram accurately to represent their journeys.

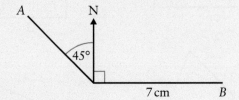

Find the final distance between the two boats.

🏆 Challenge Q9

9   Town *P* is 12 km West of town *Q*.

Town *R* is Northeast of town *P* and Northwest of town *Q*.

Draw an accurate diagram to represent their positions. Use a scale of 1 cm to represent 1 km.

Find the distance of *R* from *P* and from *Q*.

🏆 Challenge Q10

 Fact

In surveying, triangulation defines the exact position of a point by giving the angles to the point from the ends of a fixed baseline. Questions 9 and 10 are examples of triangulation.

Surveyors use triangulation to make accurate measurements.

10 Two phone masts *X* and *Y* are 5 km apart. *X* is due North of *Y*.

Another phone mast is to be built at *Z*, which is Northeast of *Y* and East of *X*.

By drawing an accurate diagram, find the distance between masts *Y* and *Z*.

⭐ Self assessment

○ I can measure angles less than 180° accurately.

○ I can draw angles less than 180° accurately.

○ I can draw and measure reflex angles accurately.

○ I can identify acute, right, obtuse and reflex angles.

○ I can name angles using angle notation.

○ I can name and use eight points of the compass.

○ I can draw a triangle using a ruler and protractor.

○ I can draw an accurate diagram to represent a journey.

I can calculate unknown angles on a straight line or at a point.

I can identify and use complementary angles.

I can identify and use supplementary angles.

I can find unknown angles and state reasons at each stage of my working.

I can draw a triangle or quadrilateral accurately from a written description.

I can identify and name scalene, isosceles and equilateral triangles.

I can identify and name acute-angled, right-angled and obtuse-angled triangles.

I can use the angle sum of a triangle and properties of triangles to find unknown angles in triangles.

I can find angles of elevation and angles of depression.

I can draw accurate diagrams to represent real-life situations and use the diagrams to find unknown lengths in real life.

? Check your knowledge questions

1 a State whether each angle is obtuse or acute.

 b Measure the size of each angle.

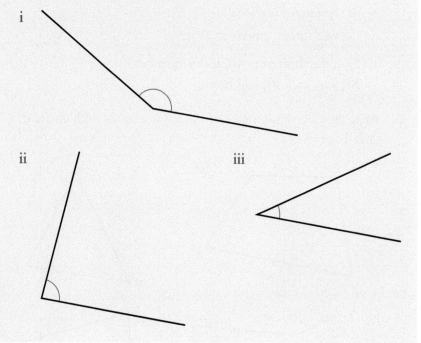

2 Measure the size of the reflex angle at vertex *D* in this shape.

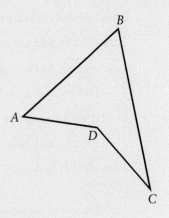

3 State the compass direction that is:

    a   180° clockwise from North   b   90° anti-clockwise from East.

4 Draw each angle:

    a   75°   b   110°   c   23°   d   330°

5 Max leaves his home and walks South for 3 km.

  Then he turns and walks Northwest until he is directly West of his home.

    a   Draw an accurate diagram on squared paper to represent Max's walk.

        Let one square represent 1 km.

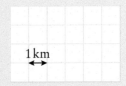

1 km

    b   Find the shortest possible distance for Max to walk straight home.

6 Write the three-letter name of the angle marked with an arc in each diagram.

  a   b

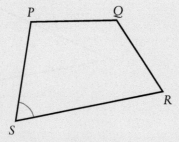 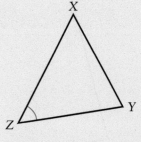

  c   d

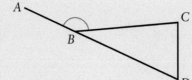 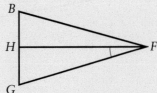

7 a In this quadrilateral,
      what type of angle is:

      i   ∢ADC

      ii ∢BCD?

   b Measure:

      i   ∢DAB

      ii ∢ABC

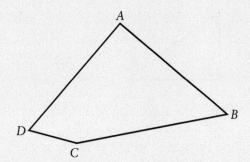

8 Work out the size of each angle labelled with a letter.

   a   b

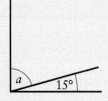   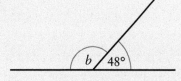

   c   d

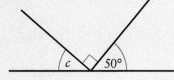   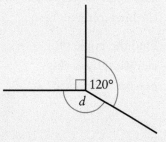

   e   f

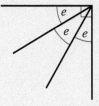   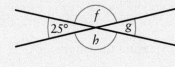

9 Write down the size of the angle that is:

   a   supplementary to 60°   b   complementary to 60°.

10 Classify each triangle.

   a   b

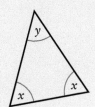   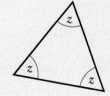

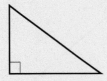

c

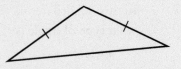

d

11 What is the angle sum of a triangle?

12 Find the size of ∢PQR in this triangle.

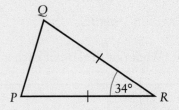

13 The diagram shows a bird in a tree and a person on the ground.

The angle of depression from the bird to the person is 55°.

Find the angle of elevation from the person to the bird.

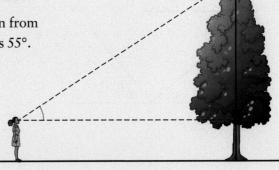

14 Draw this triangle accurately.

Measure the length *TU*.

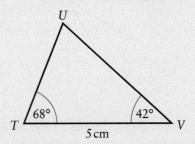

15 Draw accurately two isosceles triangles that have at least one angle of 46°.

16 Draw accurately triangle *ABC*, where *AC* = 8 cm, *BC* = 5 cm and ∢*BCA* = 35°.

Find the size of ∢*ABC*.

17 Two ships *S* and *T* are 7 km apart. *S* is due East of *T*.

Ship *R* is due South of *T* and Southwest of *S*.

By drawing an accurate diagram, find the distance between ships *R* and *T*.

Percentages and applications

5

68%

5 Percentages and applications

 KEY CONCEPT

Form

 RELATED CONCEPTS

Equivalence, Patterns, Quantity

 GLOBAL CONTEXT

Identities and relationships

Statement of inquiry

Understanding the relationship between different forms of equivalent numbers allows us to identify and use patterns so we can solve problems more easily.

Factual

- What are percentages?

Conceptual

- How do you convert between fractions, decimals and percentages?
- How do you calculate the percentage of a quantity?

Debateable

- How are percentages useful?

Do you recall?

1 How do you convert a fraction with a denominator of 10, 100 or 1000 to a decimal? For example, write $\frac{3}{100}$ as a decimal.

2 How do you simplify a fraction? For example, simplify $\frac{80}{100}$

3 How do you divide to give decimal answers? For example, work out $3 \div 4$

4 How do you change an improper fraction to a mixed number? For example, write $\frac{7}{5}$ as a mixed number.

5 How do you multiply fractions? For example, work out $\frac{3}{5} \times \frac{10}{21}$

5.1 Fractions, decimals and percentages

5.1.1 Changing fractions to decimals

You may have come across percentages before. They are commonly seen advertising discounts in shops or online. Financial institutions use percentages to quote interest charges or interest paid on savings and investments. Can you think of other examples where you have seen percentages?

Before calculating percentages of a quantity, you need to be able to convert between fractions, decimals and percentages.

When fractions are written as decimals, they are either terminating or recurring decimals.

A **terminating** decimal has a finite number of decimal places.

For example, 1.7 has one decimal place, 1.76 has two decimal places and 1.76523 has five decimal places.

A **recurring** decimal does not have a finite number of decimal places; the decimals go on forever. A recurring decimal forms a pattern that repeats.

For example, $0.4444\ldots = 0.\dot{4}$, $0.181818\ldots = 0.\dot{1}\dot{8}$, and $6.3562562\ldots = 6.3\dot{5}6\dot{2}$
To show that a digit or group of digits is repeated, a dot is placed over the first and last repeating digits.

 Fact

Finite means that it comes to an end. Infinite means that it goes on forever.

 Explore 5.1

Can you write each fraction as an equivalent fraction with 10, 100 or 1000 as the denominator?

$\dfrac{1}{2}, \dfrac{1}{4}, \dfrac{3}{4}, \dfrac{1}{5}, \dfrac{4}{5}, \dfrac{1}{20}, \dfrac{3}{20}, \dfrac{11}{20}, \dfrac{1}{200}, \dfrac{57}{200}$

Now can you write each of your fractions as a decimal?

For example, $\dfrac{1}{5} = \dfrac{2}{10} = 0.2$

All of the fractions listed above change to terminating decimals.

How would you show that $\dfrac{1}{3} = 0.\dot{3}$?

Can you find other fractions that change to recurring decimals?
Write each one you find as a recurring decimal.

Worked example 5.1

Write each fraction as a decimal.

a $\dfrac{3}{5}$ **b** $2\dfrac{3}{8}$ **c** $\dfrac{1}{6}$

Solution

a $\dfrac{3}{5} = 3 \div 5$

We can use any written method or a calculator to divide 3 by 5.
The display on a graphic display calculator (GDC) can be converted between fractions and decimals with the touch of a button. Check your calculator instructions to find the button for your model of calculator.
The GDC output will look something like this:

```
▤ Math Rad Norm1   ab/c a+bi
3÷5
                              0.6
```

So, $\dfrac{3}{5} = 0.6$

b $2\dfrac{3}{8} = 2 + 3 \div 8$

We can use any written method or a calculator to divide 3 by 8.
The GDC output will look like this:

```
▤ Math Rad Norm1   ab/c a+bi
2 3/8
                            2.375
```

So, $2\dfrac{3}{8} = 2.375$

c $\dfrac{1}{6} = 1 \div 6$

How many digits does your graphical calculator display by default?
We can choose how many decimal places a calculator displays.

```
▤ Math Rad Norm1   ab/c a+bi
1÷6
                      0.1666666667
```

So, $\dfrac{1}{6} = 0.1\dot{6}$

⇄ Reflect

Look at your answers to Explore 5.1. How could you use the decimal equivalent of $\dfrac{1}{20}$ to write $\dfrac{3}{20}$ and $\dfrac{11}{20}$ as decimals? How about $\dfrac{1}{200}$ and $\dfrac{57}{200}$?
How do you know if a fraction is a terminating or recurring decimal?
How can you tell from a calculator display if the answer is a recurring decimal?

1   Write down whether each decimal is terminating or recurring.

   a   0.75   b   0.222…   c   5.666…   d   32.8

   e $44.\dot{1}$ f   55.5   g $12.\dot{1}2\dot{3}$ h   1.204204…

   i   8.88   j   2.2   k   15.15   l $45.1\dot{6}$

2   Write each of these recurring decimals using dot notation.

   a   0.777…   b   7.525252…

   c   0.252525…   d   0.366666…

   e   9.808080…   f   3.109109…

   g   0.142857142857…   h   14.562234234…

3   Write each fraction as a decimal.

   a $\dfrac{7}{10}$ b $\dfrac{9}{10}$ c $\dfrac{3}{100}$ d $\dfrac{28}{100}$

   e $\dfrac{8}{1000}$ f $\dfrac{42}{1000}$ g $\dfrac{125}{1000}$ h $\dfrac{67}{100}$

   i $\dfrac{17}{50}$ j $3\dfrac{1}{2}$ k $\dfrac{21}{25}$ l $\dfrac{193}{500}$

4   Write each fraction as a decimal.

   a $\dfrac{2}{5}$ b $\dfrac{1}{4}$ c $\dfrac{9}{20}$ d $\dfrac{13}{50}$

   e $\dfrac{1}{8}$ f $\dfrac{5}{8}$ g $\dfrac{7}{8}$ h $\dfrac{11}{25}$

   i $\dfrac{1}{9}$ j $\dfrac{5}{9}$ k $\dfrac{1}{3}$ l $\dfrac{1}{6}$

   m $\dfrac{2}{6}$ n $\dfrac{5}{6}$ o $\dfrac{2}{3}$ p $\dfrac{7}{9}$

5   Convert these mixed numbers to decimals.

   a $2\dfrac{1}{9}$ b $7\dfrac{1}{11}$ c $5\dfrac{5}{9}$ d $12\dfrac{4}{11}$

   e $6\dfrac{1}{16}$ f $2\dfrac{5}{7}$ g $19\dfrac{8}{15}$ h $1\dfrac{17}{40}$

6   Match each decimal with an equivalent fraction.

0.4 0.24 0.35	$\dfrac{2}{5}$ $\dfrac{12}{50}$ $\dfrac{1}{2}$ $\dfrac{1}{100}$
0.01 0.3	$\dfrac{3}{10}$ $\dfrac{4}{5}$ $\dfrac{7}{20}$ $\dfrac{3}{4}$
0.8 0.5 0.75	

7   Write these numbers in ascending order.

$\frac{3}{10}, \frac{1}{3}, 0.33$

8   Zara has this list of numbers:

$0.7, \frac{3}{4}, \frac{3}{5}, 0.8$ and $\frac{2}{3}$

She says, 'The smallest number is $\frac{2}{3}$ as the digits 2 and 3 are the smallest.'

Is Zara correct? Explain how you know.

9   Write down a decimal that is greater than $\frac{2}{3}$, but less than $\frac{3}{4}$

10 Amara says, 'Multiplying by $\frac{1}{3}$ is the same as multiplying by 0.33.'

Is Amara correct? Give a reason for your answer.

11 Ayaz is asked to convert 0.35 hours into minutes.

He writes, '0.35 hours = 35 minutes'.

a   Explain why Ayaz is wrong.

b   What is 0.35 hours written in minutes?

12 Which of these fractions is closest to 0.3?

$\frac{2}{5}, \frac{5}{20}, \frac{16}{50}, \frac{1}{3}$

Explain how you know.

13 Write down a decimal that is greater than $\frac{17}{20}$, but less than $\frac{23}{25}$

14 Here is a group of fractions.

$\frac{1}{2} \quad \frac{1}{3} \quad \frac{1}{4} \quad \frac{1}{5} \quad \frac{1}{6} \quad \frac{1}{7} \quad \frac{1}{8} \quad \frac{1}{9} \quad \frac{1}{10} \quad \frac{1}{11} \quad \frac{1}{12} \quad \frac{1}{15}$

a   Sort the fractions into a set of recurring decimals and a set of terminating decimals.

b   How can you tell if a fraction converts to a recurring or terminating decimal?

Investigation 5.1

Patterns in recurring decimals

Write $\frac{1}{9}$ and $\frac{2}{9}$ as decimals.

Continue the pattern with $\frac{3}{9}, \frac{4}{9}$ and so on. Can you work these out

without using a calculator? What do you notice about the pattern?

Repeat this process with $\frac{1}{7}$ and $\frac{2}{7}$

What other recurring fractions can you find that have patterns?

5.1.2 Converting between fractions, decimals and percentages

'Per cent' means 'per 100' or 'for every 100' and is written with the % symbol. Percentages are used to write fractions with a denominator of 100.

Explore 5.2

Can you write the percentage, decimal and fraction shaded for each diagram? Give each fraction in its simplest form.

a
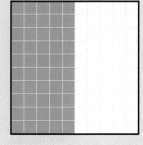

b
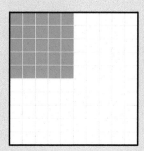

c
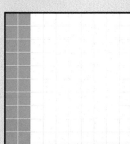

d
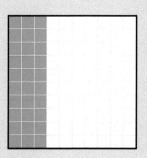

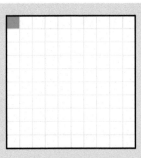

e

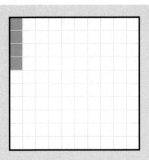

f

Using your answers, can you explain a method for converting between equivalent fractions, decimals and percentages?

 Worked example 5.2

a Write 120% as a mixed number in its simplest form.

b Write $\dfrac{4}{25}$ as a percentage.

c Write 34% as a decimal.

Solution

a $120\% = \dfrac{120}{100}$ ———— To write a percentage as a fraction or mixed number, first write it as a fraction with denominator 100.

$= 1\dfrac{1}{5}$ ———— Change the improper fraction to a mixed number and simplify.

b $\dfrac{4}{25} = \dfrac{4}{25} \times \dfrac{100}{1}\%$ ———— To write a fraction as a percentage, multiply by 100%.

$= \dfrac{4}{{}_1 \cancel{25}} \times \dfrac{\cancel{100}^4}{1}\%$ ———— Simplify the multiplication.

$= 16\%$

c $34\% = \dfrac{34}{100}$ ———— Write the percentage as a fraction with denominator 100.

$= 0.34$ ———— Now write the fraction as a decimal.

⇄ Reflect

How would you explain a method for converting a decimal to a percentage?

Can you find any other methods for converting between equivalent fractions, decimals and percentages?

Practice questions 5.1.2

1   Convert these percentages to fractions.

   a   70%   b   77%   c   7%

   d   10%   e   1%   f   29%

   g   41%   h   22%   i   33%

2   Convert each of these percentages to an equivalent fraction.
   Give each fraction in its simplest form.

   a   25%   b   20%   c   50%   d   45%

   e   65%   f   90%   g   60%   h   5%

   i   2%   j   8%

> **Hint Q2**
>
> To simplify a fraction, find the highest common factor of the numerator and denominator.

3   Convert each percentage to an equivalent decimal.

   a   21%   b   27%   c   90%   d   99%

   e   9%   f   22%   g   2%   h   1%

   i   3%   j   130%   k   129%   l   206%

4   Write these decimals and percentages in ascending order.

   30%   0.03   0.33   0.3%   3.3%

5   Convert each percentage to a whole number or mixed number in its simplest form.

   a   150%   b   325%   c   140%   d   500%

   e   440%   f   222%   g   125%   h   207%

6   Write down the equivalent percentage for each fraction or number.

   a $\dfrac{7}{100}$ b $\dfrac{12}{100}$ c $\dfrac{41}{100}$ d $\dfrac{94}{100}$

   e $\dfrac{1}{100}$ f $\dfrac{37}{100}$ g $\dfrac{1}{50}$ h $\dfrac{3}{20}$

   i $\dfrac{13}{25}$ j $\dfrac{7}{10}$ k   5   l $1\dfrac{3}{10}$

7   Match each fraction to its equivalent percentage.

| $\dfrac{3}{20}$ $\dfrac{39}{50}$ $\dfrac{3}{4}$ $\dfrac{13}{100}$
 $\dfrac{3}{5}$ $1\dfrac{1}{2}$ $\dfrac{7}{10}$ | 60%   78%
 70% 150% 75%
 13%   15% |

8   Write each of these fractions as a percentage.

   a $\dfrac{4}{5}$ b $\dfrac{11}{20}$ c $\dfrac{1}{4}$ d $\dfrac{37}{100}$

   e $\dfrac{116}{200}$ f $\dfrac{2}{3}$ g $\dfrac{1}{6}$ h $1\dfrac{3}{4}$

9   Convert these decimals to percentages.

   a   0.47   b   0.85   c   0.3   d   0.03

   e   0.33   f   0.125   g   1.7   h   3.25

   i   4.375   j   4   k   8.01   l   1.1

10   Copy and complete this table of equivalent fractions, decimals and percentages.

Fraction	Decimal	Percentage
		7%
$\dfrac{7}{20}$		
	0.7	
	0.03	
	1.25	
		40%
$\dfrac{1}{6}$		
$\dfrac{1}{3}$		
	0.09	
$\dfrac{1}{8}$		
		190%

11   Convert these percentages to fractions and decimals.

   a $2\dfrac{1}{2}\%$ b $3\dfrac{1}{4}\%$ c $4\dfrac{3}{4}\%$ d $8\dfrac{1}{10}\%$

   e   21.6%   f   13.7%   g   2.25%   h   1.07%

12 At North Manor school, $24\frac{1}{2}\%$ of the students can touch their nose with their tongue.

  a What fraction of the students can touch their nose with their tongue?

  b Express this fraction as a decimal number.

13 Three students each receive $3\frac{1}{3}\%$ of the profits from selling lemonade.

  a What fraction do they each receive?

  b The three students combine the money received. What decimal is this?

14 Which is the greatest value: $\frac{3}{5}$, 65% or 0.6?
   Explain your answer.

15 State whether each statement is correct. Give reasons for your answers.

  a $\frac{27}{50} = 54\%$ b $\frac{1}{20} = 20\%$ c $\frac{3}{10} = 3\%$

16 On an adventure course, Ibrahim has the following accidents:

  • He falls off equipment 23 times.

  • He trips over 16 times.

  • He walks into obstacles 11 times.

  a What fraction of his accidents are not due to trips?

  b What percentage of his accidents are not due to trips?

17 During an athletics competition, a team won:

  • 9 medals in sprinting events

  • 7 medals in long-distance running events

  • 3 medals in jumping events

  • 1 medal in throwing events.

  What percentage of medals were not won for jumping events?

18 Write these values in ascending order.

   $13\frac{3}{5}$ 13.5   135.5%

19 a Make a table with column headings like the one in question 10.
      Then use your table to write in the equivalent fractions, decimals and percentages from the list given.

0.5	5%	$\frac{1}{10}$	1%	0.2	$\frac{7}{50}$	15%	0.15	64%
8%	$\frac{1}{5}$	0.08	0.65	$\frac{2}{25}$	65%	0.1	14%	0.05
$\frac{3}{20}$	0.64	20%	$\frac{1}{20}$	10%	0.14	$\frac{13}{20}$	$\frac{16}{25}$	$\frac{1}{14}$

b   Which three numbers have no equivalent values in the list?

   Thinking skills

   Social skills

   Communication skills

   Connections

Game: Matching pairs

Write some equivalent pairs on blank cards, such as:

80%, $\frac{4}{5}$ 0.25, 25%   60%, 0.6   0.08, 8% $\frac{1}{10}$, 10%

1%, 0.01 $\frac{3}{20}$, 15%   0.2, 20%   2%, $\frac{1}{50}$ $\frac{3}{25}$, 12%

Shuffle your cards and spread them out, face down on a table.

Take turns to select two cards to see if you have a matching pair. If the cards form an equivalent pair, keep them and have another go. If the cards do not form an equivalent pair, turn them face down again.

The winner is the person with the most pairs when all cards have been paired up.

 5.2   **Finding a percentage of a quantity**

   Explore 5.3

Without using a calculator, can you use the grid below to find each percentage of the number 200? First work out 50% of 200, then 25% of 200, 75% of 200, and so on.

50%	20%	15%	1%	26%
25%	30%	35%	3%	99%
75%	40%	85%	4%	51%
10%	5%	95%	7%	82%

Can you explain how you can use 10% to work out 5%? How can you use 1% to work out 3%?

Can you explain any of your other calculations for working out the percentages on the grid?

Can you work out these percentages for more difficult numbers such as 240 or 136?

PER SERVING

Saturates	Calories	Salt	Fibre	Fat	Sugars
0.5g	450	0.3g	0.5g	0.5g	8g
5%	17%	3%	12%	25%	2%

of your guideline daily amount

Nutrition labels show what percentage of the recommended daily amounts of nutrients a food contains. This label shows that 8 g is 2% of the recommended amount of sugar.

Work out 12% of 350 m.

Solution

12% of 350 m = 0.12 × 350 ———— To find the percentage of a quantity, write the percentage as a decimal and multiply by the quantity.

= 42 m ———— Remember the units in your answer.

Alternative method

12% of 350 m = $\dfrac{12}{100} \times \dfrac{350}{1}$ ———— To find the percentage of a quantity, write the percentage as a fraction and multiply by the quantity.

= $\dfrac{12}{{}_4\cancel{100}} \times \dfrac{\cancel{350}^{14}}{1}$ ———— Simplify the calculation where possible.

= $\dfrac{{}^3\cancel{12}}{{}_1\cancel{4}} \times \dfrac{14}{1}$

= 42 m

 Reflect

Look at the percentages you worked out in Explore 5.3. Can you list them as calculations using decimals?

Look at both options in each of these:

* have 10% of $4 or 65% of 80 cents

* receive 45% of 2 chocolate bars or 17% of 5 chocolate bars

* be stung by 8% of 150 bees or by 12% of 50 bees

* sit a test for 40% of 2 hours or 55% of 1 hour 40 minutes.

Which option would you prefer in each case? Explain why.

 Practice questions 5.2

1 Work out the given percentage of each amount.

    a 10% of 220 g   b 30% of 400 km

    c 60% of 40 kg   d 5% of 800 g

    e 1% of 50 cm   f 3% of 30 m

    g 25% of 24 litres   h 40% of 70 kg

    i 80% of 140 km   j 75% of 14 hours

    k 90% of 90 minutes   l 9% of 300 seconds.

2 Write down these values, to the nearest cent.

    a 25% of $450   b 10% of $15

    c 6% of $300   d 2% of $400

    e 10% of $47.50   f 22% of $9.50

    g 45% of $15.60   h 7% of $6.30.

3 Ben has 30 trading cards. He loses 20% on his way home from school. How many does he lose?

4 Maisa has a worm farm containing 2000 worms.
7% of the worms are sold.
How many worms does she have left?

5 Omar has 30 hats.
90% of his hats are sports caps.
How many are not sports caps?

6 Ayemen's garden has an area of 250 m².
He wants to use 18% for growing fruit.
What area does he need to set aside for growing fruit?

7 Miriam earns $15 000 per year.
She spends 85% of this, sends 10% to her family and saves the rest.

    a How much does she spend?

    b How much does she save?

8 Suliman has the choice of taking 65% of $40 or 15% of $140.
Which should he choose? Explain your answer.

9 For Eiman to qualify as a table-football grand champion, she must win 95% of her matches.
She has played 700 matches and won 664 matches.
Does she qualify as a grand champion? Explain your answer.

10 Hakan uses this calculation to work out 8% of $40:
0.8×40
Hakan is incorrect. Explain his mistake.

 Challenge Q11

11 What is 10% of 25% of $120?

12 Two towns vote for their favourite superhero.

Town A has a population of 12 000, and 55% of people vote.

Town B has a population of 9400, and 70% of people vote.

Which town casts more votes? Explain your answer.

13 Is 20% of 25% the same as 45%?

Explain your answer.

 Challenge Q13

 Hint Q13

You could select a number to find the percentages of, then test with a different number.

14 Jameela sells a mint-condition comic book through a collector's website.

The website charges 5% to list the comic, and Jameela will have to pay $3.40 postage to send it to the buyer.

If the comic sells for $80, how much will she have left after paying for the postage and listing?

15 Three students are comparing their sticker collections.

Aisha says, '10% of my collection is 35 stickers.'

Baris says, '20% of my collection is 50 stickers.'

Camila says, '15% of my collection is 45 stickers.'

What is the difference between the number of stickers in the largest and the smallest collections? Explain your answer.

 Challenge Q15

16 Adib, Bisma and Cole enter an archery competition.

The maximum score possible is 80 points.

Adib scores 58 points.

Bisma scores 85% of the maximum points.

Cole scores $\frac{11}{16}$ of the total points.

How many points separate the highest and the lowest of the three scores?

 Connections

Game

This is a two-player game. Each player needs a different coloured pen or pencil.

Each player will need their own copy of this grid:

2.6	10.8	9.9	4	8.75
1.4	6.4	0.9	3.6	2.8
9	16.25	4.8	13.2	7.2
2.4	5	2.25	3.5	6.3
10.2	6.72	12	5.6	4.5

Player 1: Choose a number on your grid. Then choose the calculation from the grid below that you think gives the number you have selected. Check your calculation using a calculator. If you are correct, colour in the answer on your grid.

Player 2: Follow the same steps above.

5% of 80	6% of 40	85% of 12	10% of 50	15% of 6
10% of 48	7% of 80	55% of 24	14% of 10	25% of 9
12% of 90	2% of 130	65% of 25	20% of 18	20% of 45
8% of 84	15% of 30	25% of 35	80% of 8	60% of 12
4% of 70	90% of 11	75% of 16	70% of 9	50% of 7

Continue taking turns choosing a number and trying to find the correct calculation for your number.

The winner is the first player with 5 shaded numbers in a vertical, horizontal or diagonal straight line.

🔍 Explore 5.4

Can you complete the spider diagram, adding as many calculations as possible?

Add as many branches as you need for your calculations.

Can you explain the strategies you used to produce your calculations?

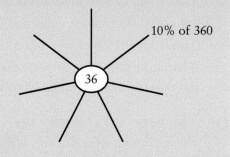

10% of 360

36

🔍 Investigation 5.2

Show that 30% of 60 = 60% of 30

Does this work for other values? For example, is 55% of 120 equal to 120% of 55?

Is this true for all values? Give reasons for your answer.

5.3 One quantity as a percentage of another

To write one quantity as a percentage of another, first write the two quantities in the same units, if they are not already. Next, write the first quantity as a fraction of the second quantity. Finally, change the fraction to a percentage.

Explore 5.5

The table shows some of Ushi's test results.

Subject	Result	Total
Language acquisition	43	50
Language and literature	64	80
Individuals and societies	8	10
Sciences	50	60
Mathematics	113	120
Arts	17	20
Physical and health education	38	40
Design	22	25

Can you find the subject in which Ushi performed the best?

Worked example 5.4

a Noel collects comics. He has 80 comics about one superhero.

The superhero uses all of his superpowers in 50 of the 80 comics. In what percentage of these comics are all of this superhero's superpowers used?

b Work out 56 cm as a percentage of 2 m.

Solution

a **Understand the problem**

We need to work out 50 as a percentage of 80.

Make a plan

First we will write 50 out of 80 as a fraction. Then we will convert the fraction to a percentage.

Carry out the plan

$$\frac{50}{80} = \frac{50}{80} \times \frac{100}{1}\%$$ — Multiply by 100%.

$$= \frac{50}{\cancel{80}_4} \times \frac{\cancel{100}^5}{1}\%$$ — Simplify the multiplication.

$$= \frac{\cancel{50}^{25}}{\cancel{4}_2} \times \frac{5}{1}\%$$

$$= \frac{125}{2}\%$$

$$= 62\frac{1}{2}\% \text{ or } 62.5\%$$

Look back

We can work backwards to check if the solution is correct:

62.5% of 80 = 0.625 × 80 = 50

b **Understand the problem**

We need to work out 56 cm as a percentage of 2 m.

Make a plan

First we will write the two quantities in the same units as a fraction.
Then we will convert the fraction to a percentage.

Carry out the plan

2 m = 200 cm

So, the two quantities are 56 cm and 200 cm.

Writing the two quantities as a fraction: $\frac{56}{200}$

$$\frac{56}{200} = \frac{56}{200} \times \frac{100}{1}\% \quad \text{———— Multiply by 100\%.}$$

$$= \frac{56}{{}_{2}200} \times \frac{100^{1}}{1}\% \quad \text{———— Simplify the multiplication.}$$

$$= \frac{56}{2} \times 1\%$$

$$= 28\%$$

Look back

We can work backwards to check if the solution is correct:

28% of 200 = 0.28 × 200 = 56

 Reflect

Can you think of another method for writing one quantity as a percentage of another?

 Practice questions 5.3

1   Write down:

   a   25 cm as a percentage of 1 m

   b   400 g as a percentage of 1 kg

    c   84 marks as a percentage of 100 marks

    d   60 g as a percentage of 300 g

    e   140 km as a percentage of 400 km

    f   130 tennis balls as a percentage of 200 tennis balls.

2   Express each quantity as a percentage.

    a   500 m of 1 km   b   90° of one full turn

    c   30 years of one century   d   16 of 200

    e   15 minutes of one hour   f   70 mm of 20 cm

    g   450 of 9000   h   36 minutes of 1 hour.

3   At a water park one day, there are 24 members of staff.

    Visiting the park are 84 adults, 72 boys and 60 girls.

    Find the percentage of the people at the water park who are:

    a   adult visitors   b   members of staff

    c   boys   d   boys or girls

    e   not members of staff   f   not girls.

4   Hana collected 1200 eggs last month.

    264 of the eggs were white.

    What percentage of her eggs were white?

5   Parsa's team won 24 of their 40 matches last season.

    What percentage did they win?

6   Of the 20 computer games Dina owns, she has completed 7.

    What percentage of games has she completed?

7   Raed was absent for 40 days of school last year.

    Five of these absences were because he fell off his skateboard.

    What percentage of his absences were due to his skateboard accidents?

8   Raed paid $120 for his skateboard and some time later he sold it for $45.

    What percentage of the cost did he receive?

9   A paintball team won 13 games, drew 8 games and lost 7 games.
Find the percentage of games that:

a   they won   b   they did not lose.

10 Captain Zaarg, of the Intergalactic Peacekeepers, is to travel
60 000 000 km from Mars to Earth.

The spaceship is damaged by an asteroid after travelling 12 000 000 km.

What percentage of the journey is left to complete?

11 Nabila's ice lolly is 8 cm long when she leaves it in the sun.
When she returns it is only 68 mm long.

What percentage of the length has melted?

12 To avoid retaking a test, Noha must score 70 % or more.

She scores 22 out of a possible 30 points on the first attempt.

Will she have to retake the test? Explain your answer.

13 Liam makes 5 litres of fruit juice.

He fills five 180 ml cups for his friends and another for himself.

What percentage of the juice is left?

14 Which is the highest percentage, $\frac{17}{20}$, $\frac{21}{25}$ or $\frac{26}{30}$?

15 Paris enters a gaming competition.
There are three rounds, with 75 points available in each round.
Paris scores:
- 65 points in the first round
- 39 points in the second round
- 58 points in the third round.

The three scores are combined at the end of the third round and
players who exceed 70 % go through to the final round.

Does Paris go through to the final round? Explain your answer.

Communication skills

16 Mazin makes a fruit drink with 2 litres of water and 250 ml of fruit
juice.

Ola makes a fruit drink with 1.5 litres of water and 180 ml of fruit
juice.

Who has the higher percentage of fruit juice in their drink?

Explain your answer.

5.4 **Probability**

Probability is the chance of an event happening. An event can be a mathematical event or a situation such as finding your keys under the sofa, the chance of it raining while you are out or the chance of a machine breaking down in a factory. We can use words such as impossible, unlikely, even-chance, likely and certain to describe probability.

Weather forecasts use probability to predict the weather based on what has happened in the past.

 Explore 5.6

Flip a coin 20 times and record whether it lands on heads or tails each time. How many heads and how many tails did you get? Is this the result you expected?

Can you write some events that have a probability of $\frac{1}{2}$ or 50% that they will happen?

For example, the chance of randomly picking a black sock out of a drawer containing 8 black socks and 8 blue socks, or the probability that this spinner will land on red (R).

Probabilities can be shown on a probability scale such as:

Impossible   Even-chance   Certain

Where should your events go on the probability scale?

Roll a dice 30 times and record the number it lands on for each roll. How many of each number did you get? Is this the result you expected? Where will the probability of scoring a 1 go on the probability scale? Is this the same for each outcome on a dice?

 Worked example 5.5

a   List all possible outcomes for rolling a six-sided dice.

b   Write down the probability of this spinner landing on red (R). Give your answer as a percentage.

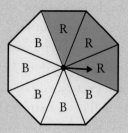

c Emily has a box of chocolates.

7 of the chocolates are milk chocolate, 3 are dark chocolate and 2 are white chocolate.

Emily selects a chocolate at random.

Write down the probability that it is a milk or a dark chocolate.
Write your probability as a fraction in its simplest form.

Solution

a All possible outcomes are: 1, 2, 3, 4, 5, 6

b 3 sections out of a total of 8 sections are red, so the probability of it landing on red is $\frac{3}{8}$

$\frac{3}{8} \times \frac{100}{1}\%$ ⟵————— Multiply by 100%.

$= \frac{3}{{}_2 8} \times \frac{\cancel{100}^{25}}{1}\%$ ⟵————— Simplify the multiplication.

$= \frac{75}{2}\%$

$= 37\frac{1}{2}$ or 37.5%

c We need to work out how many of the chocolates are milk or dark chocolates and work out how many chocolates there are altogether.

10 of the chocolates are either milk or dark chocolate.

There are 12 chocolates altogether.

The probability of Emily selecting a milk or a dark chocolate is $\frac{10}{12}$

$\frac{10}{12}$ simplifies to $\frac{5}{6}$

We can work backwards to check our answer. $\frac{5}{6}$ of 12 is 10.

 Reflect

Decide whether you agree or disagree with each statement. Give reasons for your decisions.

* It always rains in July.
* 1 is the easiest number to roll on a dice.
* I will see someone I know when I next go shopping.
* I will exercise today.
* We will have mathematics tomorrow.

1 From the list below, choose a description that best fits each situation given.

| Impossible | Unlikely | Even-chance | Likely | Certain |

   a You will roll an even number when you roll a dice.

   b You will walk on pink grass.

   c Your teacher will give you homework today.

   d The sun will rise tomorrow.

   e You will eat an ice-cream today.

   f You will eat a meal with your family today.

   g You will roll a number less than 7 when you roll a dice.

   h You will meet a superhero this week.

2 Copy the probability scale below.

   Impossible   Even-chance   Certain

   Using the events from question 1, place the letters from **a** to **h** along the scale where you think the events would appear.

3 A fair, six-sided dice is rolled. Write down the probability that the dice lands on:

   a 1   b 2

   c an odd number   d a number less than 4

   e a number greater than 4   f 0.

4 Write the probability of selecting the counter given from each bag. Give your answers as fractions.

   a black   b black

c   black

d   white

e   black

f   white.

5   Write each probability from question 4 as a percentage.

6   Draw a bag of counters where the probability of selecting a white counter at random is four times the probability of selecting a black counter.

7   a   Using this spinner, write down the probability, as a fraction, that the spinner stops on:

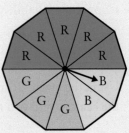

   i   green (G)
   ii   blue (B)
   iii red (R).

   b   What is the total of all three probabilities?

8   Copy this probability scale.

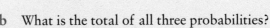

   0 $\frac{1}{2}$ 1

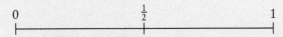

   Place the labels **i** to **iii** so that the probabilities from question **7a** are in the correct position on the line.

 Connections

Look back at Chapter 3 to remind yourself about square numbers, prime numbers and composite numbers.

9   This 12-sided dice is rolled.

   Write down the probability that it lands on:

   a   an even number

   b   a square number

   c   a prime number

   d   a composite number

   e   neither a composite nor a prime number.

10 Copy this 8-sided spinner.
Label the sections with the numbers 1, 2, 3 and 4 so that the numbers satisfy these conditions:

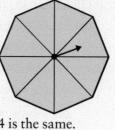

- The spinner is more likely to land on 3 than any other number.
- The probability that the spinner lands on 2 or 4 is the same.
- The spinner is less likely to land on 1 than any other number.

11 Hatem has 28 pet snakes.
12 of the snakes are poisonous.
A snake is chosen at random.
Write down the probability that the snake is not poisonous.

12 Jasmine has a collection of football shirts.
4 shirts are for Spanish teams.
5 are for German teams.
7 are for Italian teams.
A football shirt is selected at random.
Write down the probability that it is for a German team.

13 Look at these bags containing red (R), blue (B) and green (G) counters.

🛡 **Hint Q13**

P(R) means the probability of selecting a red counter.

Bag A   Bag B   Bag C   Bag D   Bag E   Bag F

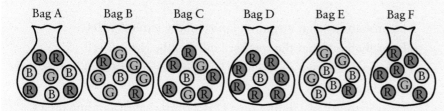

$P(R) = \frac{1}{8}$ $P(R) = \frac{7}{8}$ $P(R) = \frac{1}{4}$ $P(R) = \frac{1}{2}$ $P(R) = \frac{5}{8}$

a Match each bag of counters to its corresponding probability. There will be one bag left over.

b For the bag that is left over, write the probability of selecting a red counter.

Raul and Yousef play a game. Raul wins if he picks a red counter and Yousef wins if he picks a green counter. The boys can select between bags A to F.

c Which bag should Raul choose to play the game? Explain your answer.

d Which bag should Yousef choose to play the game? Give a reason for your answer.

14 Olivia has 30 comics in her collection. She has 13 Captain Australia comics, 8 Wonder Man comics and the rest are Purple Blob comics.
Olivia selects a comic at random.
Find the probability that she: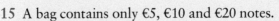

   a   selects a Purple Blob comic

   b   does not select a Wonder Man comic

   c   selects a Moth Woman comic.

 Hint Q15

P(€1) means 'the probability of €1'.

15 A bag contains only €5, €10 and €20 notes.
There are four times as many €10 notes as €20 notes.
There are twice as many €5 notes as €10 notes.
Write down:

   a   P(€5)   b   P(not €10)

   c   the smallest possible value for the total amount of money in the bag.

 **Challenge Q16**

16 Use the digits 1, 2, 3, 4 and 5 to form a set of numbers that satisfy these conditions:

$$P(\text{not } 5) = \frac{2}{5} \text{ and } P(1) = \frac{1}{5}$$

You do not have to use every digit, and you may use a digit more than once.

 **Challenge Q17**

17 The probability that a spinner lands on the letter A is 0.05.
The probability that the spinner lands on the letter B is 0.15.
The probability that the spinner lands on the letter C is 12 times that of landing on an A.
The probability of landing on a D is a third that of landing on a C.
Write down the probability that the spinner lands on an E.
Give a reason for your answer.

18 Look at these bags of counters.

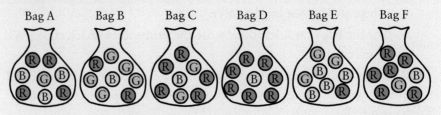

$$P(\text{not } G) = 1 \quad P(B \text{ or } G) = \frac{3}{4} \quad P(B \text{ or } G) = \frac{1}{4} \quad P(R \text{ or } B) = \frac{7}{8} \quad P(\text{not } R) = \frac{7}{8}$$

Match each bag of counters to its corresponding probability so that there is one bag left over.

Explore 5.7

A two-player game involves rolling one dice.
If the dice lands on a prime number, player 1 wins.
If the dice lands on a 6, player 2 wins.
Can you decide who is most likely to win this game? Explain why.
Can you think of alternative rules to make this a fair game?

 Thinking skills

 Social skills

 Communication skills

Self assessment

I can recognise terminating and recurring decimals.

I can convert fractions and mixed numbers to their equivalent decimals.

I can convert between equivalent fractions, decimals and percentages.

I can work out a percentage of a quantity.

I can write one quantity as a percentage of another.

I can compare quantities using percentages.

I can use words to describe probabilities.

I can use fractions and percentages to write probabilities.

? Check your knowledge questions

1 Write down whether each decimal is terminating or recurring.

   a 3.2555...   b 27.27   c 0.5̇34̇   d 478.1̇

2 Write each of these recurring decimals using dot notation.

   a 0.555...   b 3.646464...   c 0.17777...   d 2.516516...

3 Write each of these fractions as decimals.

   a $\frac{9}{100}$ b $\frac{64}{100}$ c $\frac{175}{1000}$ d $\frac{17}{50}$ e $4\frac{1}{2}$ f $\frac{12}{25}$

   g $\frac{13}{20}$ h $\frac{3}{5}$ i $\frac{3}{8}$ j $\frac{7}{9}$ k $\frac{1}{7}$ l $\frac{1}{11}$

4 Convert these mixed numbers to decimals.

   a $2\frac{1}{3}$ b $2\frac{3}{16}$ c $5\frac{4}{7}$ d $12\frac{4}{15}$

5 Write these numbers in ascending order.

   $\frac{7}{10}$ 0.65 $\frac{2}{3}$ 0.6

6 Write a decimal that is greater than $\frac{1}{3}$, but less than $\frac{2}{5}$

7 Which of these fractions is closest to 0.5?

   $\frac{3}{5}$ $\frac{9}{20}$ $\frac{23}{50}$ $\frac{4}{7}$

   Explain how you know.

8 Convert each percentage to an equivalent fraction in its simplest form.

a 40%   b 35%   c 95%   d 4%

9 Convert each percentage to an equivalent decimal.

a 37%   b 30%   c 96%   d 4%   e 180%   f 103%

10 Write these decimals and percentages in ascending order.

70%   0.07   0.77   0.7%   7.7%

11 Convert each percentage to a whole number or mixed number in its simplest form.

a 250%   b 175%   c 300%   d 324%

12 Write down the equivalent percentage for each fraction.

a $\dfrac{3}{100}$ b $\dfrac{7}{50}$ c $\dfrac{13}{20}$ d $\dfrac{16}{25}$

e $\dfrac{3}{10}$ f $1\dfrac{7}{10}$ g $\dfrac{124}{200}$ h $\dfrac{5}{6}$

13 Convert these decimals to percentages.

a 0.23   b 0.4   c 0.01   d 0.375   e 1.5   f 2.07

14 Copy and complete this table of equivalent fractions, decimals and percentages.

Fraction	Decimal	Percentage
		4%
$\dfrac{17}{20}$		
	0.1	
	0.06	
	2.75	
		60%
$\dfrac{5}{6}$		
$\dfrac{2}{3}$		
$\dfrac{5}{8}$		
		130%

15 Which is the greatest value: $\frac{7}{20}$, 33% or 0.3?

   Explain your answer.

16 In a mathematics test, Joe scores 80 marks.

   He scores 32 marks on number.

   He scores 23 marks on geometry.

   He scores 25 marks on algebra.

   a   What fraction of his marks are not on number?

   b   What percentage of his marks are not on number?

17 Write down these values in ascending order.

   $24\frac{4}{5}$ 24.4   224.5%

18 Work out:

   a   40% of 60 kg   b   15% of 500 ml

   c   7% of 40 m   d   25% of 10 hours.

19 Write down these values to the nearest cent.

   a   10% of $12   b   8% of $500

   c   10% of $19.50   d   44% of $7.50.

20 In a school of 350 students, 6% have blue eyes.

   How many students have blue eyes?

21 Aaron, Binita and Callum sit a test.

   The test is out of 40.

   Aaron scores 32.

   Binita scores 85%.

   Callum scores $\frac{7}{8}$ of the total marks.

   Who achieved the highest marks in this test?

   Give reasons for your answer.

22 Express each quantity as a percentage.

   a   75 cm as a percentage of 1 m

   b   600 g as a percentage of 1 kg.

23 A team competing at the Olympics won 60 medals. 36 of them were gold medals.

What percentage of the medals the team won were gold medals?

24 Zainub sold her phone for $225. It cost her $720.

What percentage of the cost did she receive?

25 A six-sided dice is rolled. Write down the probability that the dice lands on:

a   4   b   a prime number

c   a number less than 3   d   7.

26

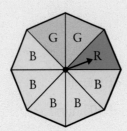

Write down the probability that the spinner stops on:

a   green (G)   b   red (R)   c   blue (B).

27 Copy this probability scale.

```
0                          1/2                        1
|---------------------------|-------------------------|
```

Place the letters **a** to **c** so that the probabilities from question 26 are in the correct position on the line.

28 A bag contains red, blue and green counters only.

There are three times as many red counters as green counters.

There are twice as many blue counters as green counters.

Write down:

a   P(red)

b   P(not green)

c   the smallest possible number of counters in the bag.

Patterns and rules

6

6 Patterns and rules

 KEY CONCEPT

Form

 RELATED CONCEPTS

Generalisation, Models, Patterns

 GLOBAL CONTEXT

Identities and relationships

Statement of inquiry

Identifying patterns presented in different forms and using a system of generalisation to communicate what we can see helps us to understand relationships and make predictions.

Inquiry questions

Factual
- If two patterns have common terms, can they still differ?
- What is a variable?

Conceptual
- How can the future behaviour of a phenomenon be predicted from present knowledge?

Debateable
- What are the advantages of algebra over other representations of phenomena?

Do you recall?

1   a   What is the order in which the operations of addition, subtraction, multiplication and division must be performed?

    b   Evaluate $5 + 4 \times 2 - 6 \div 3$

2   a   How do we show that the operations need to be performed in a different order?

    b   What is the difference between $3 + 4 \times 2$ and $(3 + 4) \times 2$?

3   a   How are numbers placed on a number line?

    b   Place the numbers $-3, 0.5, 1$ and $\frac{3}{2}$ on the number line.

6.1 Patterns

Patterns in real-life phenomena, such as the weather, allow us to predict what will happen in the future. When we can spot patterns, we can make more informed and better decisions.

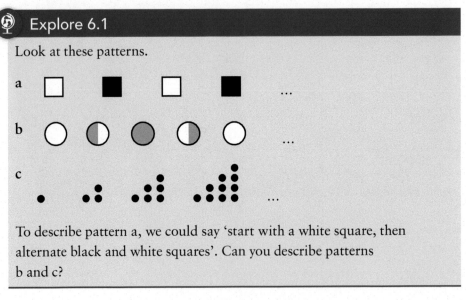

Explore 6.1

Look at these patterns.

a

b

c

To describe pattern a, we could say 'start with a white square, then alternate black and white squares'. Can you describe patterns b and c?

Each example in Explore 6.1 has a rule that makes it possible to predict the next figure in the pattern. To help us describe the rule that makes a pattern, we can label each figure with a number that tells us the position of the figure in the pattern. For example, consider the following pattern.

We can label the pattern like this:

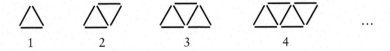

This labelling helps us to describe the pattern with more detail and precision. We can say 'figure 1 in the pattern is a triangle' or 'figure 1 is made of three segments'. We can also investigate the pattern more efficiently.

Worked example 6.1

How many line segments are needed to draw figure 15 in the triangle pattern on the previous page?

Solution

Understand the problem

The pattern is made up of line segments that form triangles. We need to find the number of line segments that form figure 15 in the pattern.

Make a plan

Method 1 We can draw all the figures from 1 to 15, and then count the line segments in figure 15.

Method 2 We can work out how many line segments are added to make each new figure, then work out how many we would need to add to make figure 15.

Carry out the plan

Method 1 After drawing all the figures from 1 to 15, figure 15 looks like this

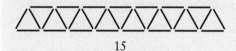

15

There are 31 line segments.

Method 2 Figure 1 has three line segments. To draw figure 2 we need to add 2 line segments to figure 1.

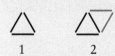

1   2

To draw figure 3, we need to add another 2 line segments to figure 2, and so on. We can organise this information in a table.

Figure number	1	2	3	4	5	...
Number of line segments	3	3 + 2	5 + 2	7 + 2	9 + 2	...

Now we have changed the pattern of figures into a sequence of numbers. A number sequence is an ordered list, with numbers separated by commas:

3, 5, 7, 9, 11, 13, ...

Now we need to find the 15th number in the sequence.

Continuing the number pattern gives:

3, 5, 7, 9, 11, 13, 15, 17, 19, 21, 23, 25, 27, 29, 31

There are 31 line segments in figure 15.

Look back

Is there a different way to calculate the number of line segments needed?

 Hint

It is often convenient to use tables to display patterns.

 Fact

The three dots at the end of the sequence are called an ellipsis. The word ellipsis originates from the Ancient Greek: Ἔλλειψις, élleipsis, meaning 'leave out'. In mathematics, it means 'continue in the same way' or 'continue this pattern'.

Reflect

Which method in Worked example 6.1 do you prefer? What are the advantages and disadvantages of drawing all the figures? What are the advantages and disadvantages of changing the pattern into a number sequence?

Which method would you use to work out how many segments are needed to draw figure 100?

In Worked example 6.1, we were able to work out the rule that relates the figures in the pattern. In this case, the rule was 'Start with 3, and add 2 each time to get the next number.' In the next sections, we will learn how to describe these rules using **algebra**.

Explore 6.2

Sketch your own pattern. Start with a simple figure, establish a rule, and repeat it to create the next figure. Make sure you are able to describe the rule.

Describe your pattern in words so that other students can draw at least the first three figures.

Patterns and rules are often used to describe how a process evolves over time. In the next example, the position reveals at what time the figure occurred in the pattern.

Worked example 6.2

A water plant grows in a circular pond. The surface of the pond that the plant covers doubles every day.

If the plant covers half of the pond after 30 days, when will it cover the whole pond?

Solution

We need to find the number of days the water plant takes to cover the whole pond.

We can draw a sketch to help us understand the process. We do not know how much of the pond was covered on day 1, but we do know that the covered surface is twice as big on day 2. We also know that half of the pond is covered on day 30.

Day 1

Day 2

...

Day 30

We can represent what we know so far in a table:

Number of days	1	2	3	...	29	30	31	...
Fraction of surface covered				...		$\frac{1}{2}$		...

So far, the only number we know in the second row is $\frac{1}{2}$ on day 30. We can use this to work out other numbers in the second row. If the plant's surface doubles every day, this means that on the 29th day the fraction covered is $\frac{1}{4}$, because this is the number that, when doubled, gives $\frac{1}{2}$. We need to find the day in which the number in the second row is 1, because this is when the whole pond is covered.

...

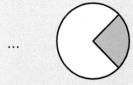

Day 29

Day 30

...

Day ?

If we double $\frac{1}{2}$ we get 1, so it takes just one more day to cover the pond. The pond will be fully covered after 31 days.

With some experience it is possible to see the answer to Worked example 6.2 at a glance. Now that you have seen the answer, can you extend your new understanding to similar problems? Are there related questions that you can answer now?

For example, can you complete the table now? What fraction of the pond was covered on day 1? Does it make sense to continue the pattern beyond day 31?

Connections

How might knowledge of this pattern benefit a biologist studying the pond and its ecosystem? Do you think it is desirable to let the plant grow to cover the whole pond? Knowledge of how fast non-indigenous plants spread is crucial for botanists who study and maintain fragile ecosystems such as rainforests, coral reefs and the Florida Everglades.

Scientists studying coral bleaching need to understand how fast it spreads.

Practice questions 6.1

1 On plain paper, draw the patterns below. Then draw the next two figures for each pattern. In your own words, state the rule that describes each pattern.

a
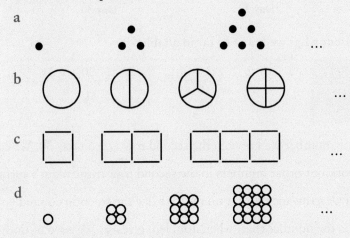

b

c

d

2 For the patterns in question 1, copy and fill in the tables with the missing numbers.

a

Position in pattern	1	2	3	4	5
Number of dots	1	3			

b i

Position in pattern	1	2	3	4	5
Number of slices	1	2	3		

ii

Position in pattern	1	2	3	4	5
Size of slice	1	$\frac{1}{2}$	$\frac{1}{3}$		

c i

Position in pattern	1	2	3	4	5
Number of squares	1				

ii

Position in pattern	1	2	3	4	5
Number of line segments	4				

d

Position in pattern	1	2	3	4	5
Number of circles					

3 Look at this diagram of boxes stacked in a shop.

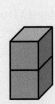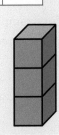

a How many square faces are visible from this point of view for:

   i one box   ii two boxes   iii three boxes?

b Copy and complete the table.

Number of boxes	1	2	3	4	5	6	7
Number of visible faces							

c Describe the pattern in words.

6.2 Number patterns

Explore 6.3

Can you find the next two numbers in each pattern?

a 7, 14, 21, 28, ...

b 0, 1, 0, −1, 0, 1, 0, −1, ...

c 1, 2, 4, 8, 16, ...

d 128, 64, 32, ...

How simply could you describe the rule for each pattern? What maths words do you need for your descriptions?

Which patterns would be easiest to draw? Are there any patterns that would be very difficult or impossible to draw?

In this section, we will learn about number patterns and the rules that describe them. Just like with figures and shapes, numbers can form patterns. For each pattern in Explore 6.3, there is a rule that relates each number to the previous number, or numbers.

For number patterns, the rule can often be described in terms of the mathematical operations of addition, subtraction, multiplication and division.

Using a rule is a convenient way of describing a pattern. The rule can be used to generate all numbers in the pattern.

Each element in a sequence is called a **term**.

Sometimes the rule for a pattern is not given. You have to work out the rule from the information you have and then use it to continue the pattern.

If you have ever played hopscotch, or similar playground games, you have been using number patterns without knowing it.

Worked example 6.3

a The rule for a number pattern is 'Start with 4, then add 15 to find the next number.'
Find the next two numbers in the pattern.

b The rule for a number pattern is 'Start with 100, then divide by 2 to find the next number.'
Find the next two numbers in the pattern.

c The rule for a number pattern is 'Start with 5. Continue the pattern by multiplying by 3 and subtracting 4 each time.'
Find the next two numbers in the pattern.

Solution

a Apply the rule to find the continuation of the pattern 4, ?, ?, ...
The second number will be $4 + 15 = 19$
The third number will be $19 + 15 = 34$
The pattern is 4, 19, 34, ...

b Apply the rule to find the continuation of the pattern 100, ?, ?, ...
The second number will be $100 \div 2 = 50$
The third number will be $50 \div 2 = 25$
The pattern is 100, 50, 25, ...

c Again, apply the rule to find the continuation of the pattern 5, ?, ?, ...
The second number will be $5 \times 3 - 4 = 11$
The third number will be $11 \times 3 - 4 = 29$
The pattern is 5, 11, 29, ...

 Reflect

Is the order in which the operations are performed important? Do the numbers in the pattern change if we change the order of the operations?

If the rule in Worked example 6.3, part c, was 'Start with 5. Subtract 4 and then multiply by 3 each time.', how would the pattern change?

Worked example 6.4

For each number pattern, state the rule in words and find the next three terms.

a 4, 10, 16, ...

b 4, 10, 25, ...

Solution

When working out the rule for a pattern, there are two common rules to check:

- Are we always adding or subtracting the same number?
- Are we always multiplying or dividing by the same number?

a We can start by checking if the difference between two consecutive terms in the pattern is always the same.

$$10 - 4 = 6$$

$$16 - 10 = 6$$

The two differences are both 6, so the rule is 'Start with 4 and add 6 each time.' This is an example of an arithmetic pattern.

To find the next three terms, we apply the rule we just found.

$$16 + 6 = 22$$

$$22 + 6 = 28$$

$$28 + 6 = 34$$

The terms in the pattern are 4, 10, 16, 22, 28, 34, …

b Again, we can start by checking if the difference between two consecutive terms is the same.

$$10 - 4 = 6$$

$$25 - 10 = 15$$

The differences are not the same. Next, we can check if the ratio between two consecutive terms is always the same.

$$\frac{10}{4} = 2.5$$

$$\frac{25}{10} = 2.5$$

Both ratios are 2.5, so the rule for this pattern is 'Start with 4 and multiply by 2.5 each time.' This is an example of a geometric pattern.

To find the next three terms, we apply the rule we just found.

$$25 \times 2.5 = 62.5$$

$$62.5 \times 2.5 = 156.25$$

$$156.25 \times 2.5 = 390.625$$

The terms in the pattern are 4, 10, 25, 62.5, 156.25, 390.625, …

 Fact

If we always add or subtract the same number, the pattern is an **arithmetic** pattern or sequence. If we always multiply or divide by the same number, the pattern is a **geometric** pattern.

If you play each of the C notes on a piano keyboard in turn, you are playing a geometric pattern. Each note has a frequency, measured in hertz, and the frequency of each C note is double the frequency of the C note before it.

Reflect

In Worked example 6.4, both patterns start with the same two terms, 4 and 10. Can we be sure that we have found the right pattern when we have found a rule based on only a few numbers?

Look again at the pattern in part b. Can we find another rule in this number pattern? How many decimal figures does each number in the pattern have? Does this pattern continue?

If we want to describe a pattern as completely as possible, which is better: stating the first hundred (or thousand) numbers in the pattern, or stating the rule in words? Why?

It is often useful to use rules in practical situations.

 Worked example 6.5

A mobile-phone company charges its customers $3.00 per month for a contract. In addition to this fixed cost, each gigabyte of data traffic used costs $0.50.

a In October, Julia used 6 gigabytes. How much did she spend in October?

b In November, Julia was billed $7.50. How many gigabytes did she use?

Solution

The question gives us a rule to find the cost based on the number of gigabytes used. We can make a table relating the cost to the number of gigabytes.

To answer part a, we will look for 6 in the gigabytes row and read off the corresponding cost.

To answer part b, we will look for 7.50 in the cost row and read off the corresponding number of gigabytes.

This also tells us where to stop in the table: we only need to get to 6 gigabytes and $7.50.

Number of gigabytes used	0	1	2	3	4	5	6	7	8	9
Cost ($)	3.00	3.50	4.00	4.50	5.00	5.50	6.00	6.50	7.00	7.50

a To find the cost of 6 gigabytes, find the entry in the table that has 6 in the first row. The corresponding cost is $6.00, so Julia was billed $6.00 in October.

b To find the number of gigabytes used when the cost is $7.50, find the entry that has 7.50 in the second row.

The corresponding number of gigabytes is 9, so Julia used 9 gigabytes in November.

We can express in words the rule that relates gigabytes used and cost in dollars as, 'Start with $3.00 and add $0.50 for each gigabyte used.'

🔍 Investigation 6.1

Two consecutive terms in a pattern are 2 and 10.

Find as many single-operation rules as you can that fit the pattern.

Find as many two-operation rules as you can that fit the pattern.

◈ Fact

A single-operation rule contains only one operation (+, −, ×, ÷), for example, 'Start with 7, **add** 12 each time.'

A two-operation rule contains two operations, for example, 'Start with 7, **multiply** by 2 and then **add** 12 each time.'

✏️ Practice questions 6.2

1 Apply the given rule to write down the first four terms of each pattern.

    a   Start with 5, add 3.

    b   Start with 10, subtract 2.

    c   Start with 1, multiply by 3.

    d   Start with 25, divide by 5.

    e   Start with −10, add 9.

    f   Start with 5, subtract 3.

    g   Start with 2, multiply by −1.

    h   Start with 9, divide by −3.

    i   Start with 1, multiply by 3 then add 1.

    j   Start with 1, add 1 then multiply by 3.

    k   Start with 3, multiply by 2 then subtract 4.

    l   Start with 2, square the number.

2 Write the rule for each pattern in words, then write down the next two terms.

    a   7, 14, 21, …   b   −1, −4, −7, …

    c   4, 20, 100, …   d   2, 0, −2, …

    e   10, 5, 2.5, …   f   4, −4, 4, …

    g   5, 50, 500, …   h   2, 4, 6, …

    i   2, 4, 8, …   j   2, 4, 16, …

    k $-\dfrac{1}{2}, \dfrac{1}{2}, \dfrac{3}{2}, \ldots$ l $1, \dfrac{1}{3}, \dfrac{1}{9}, \ldots$

3 Write down the first two terms in each pattern.

a ___ , ___ , 5, 10, 15, ...   b ___ , ___ , −1, −2, −3, ...

c ___ , ___ , 5, 25, 125, ...   d ___ , ___ , 100, 10, 1, ...

e ___ , ___ , 2, 2.5, 3, ...   f ___ , ___ , $\dfrac{1}{4}, \dfrac{1}{8}, \dfrac{1}{16}$, ...

g ___ , ___ , 15, 8, 1, ...

6.3 Variables and rules

6.3.1 Explicit rules

 Fact

Rules that relate each term in a pattern to the previous term (or terms) are called **recursive** rules. Recursive rules are often used in computer programs.

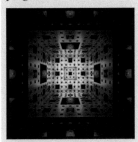

Recursive programming can be used to produce fascinating images such as this Menger sponge.

🎙 **Explore 6.4**

The rule for a pattern is 'Start with 3 and add 5.'

Can you work out the first four terms of the pattern?

Can you calculate the tenth number in the pattern without knowing the ninth? What do you need to do to calculate the ninth number?

How would you find the 500th term?

Now look at the pattern 3, 6, 9, ... How would you find the 500th term in this pattern? Can you think of a quick method?

In the patterns we have seen so far, to find a particular term we need to find all the previous terms in the pattern. It is often more convenient to find a rule for the pattern that can be expressed in terms of the *position* of the *term in the pattern*. This is called an **explicit rule**.

 Worked example 6.6

Calculate the 500th term in the pattern 4, 8, 12, ...

Solution

The difference between consecutive terms in this pattern is 4, so the rule is 'Start with 4, then add 4 each time.'

The first few terms are 4, 8, 12, 16, 20, ... and so on. We could find the 500th term by continuing to add 4 each time, but we would have to perform another 495 additions. Even if we were very fast at adding, this process would take a lot of time!

We can make a plan to answer the question in a faster way. Start by placing the terms in the pattern in a table, as we did in Section 6.1. The first row will contain the position numbers: 1, 2, 3, and so on.

Term position	1	2	3	4	5	...
Term value	4	8	12	16	20	...

Now we can look at the term values in two different ways. So far, we have always related each term to the term on its left:

Term position	1	2	3
Term value	4	8	12

add 4   add 4

Instead, we can try to relate each term to the number above it in the table, which is its position number:

Term position	1	multiply by 4	2	multiply by 4	3	multiply by 4
Term value	4		8		12	

Each term in the pattern is equal to its **position** number multiplied by 4. We can change our rule to 'The value of each term in the pattern is 4 multiplied by the term's position.'

Now we can use this new rule to find the value of the 500th term.

The 500th term is in position 500, so its value is $4 \times 500 = 2000$

Term position	1	2	...	499	500	501	...
Term value	4	8	...	1996	2000	2004	...

Using this approach, we do not need to calculate the previous numbers in the pattern to answer the question. We can quickly find any term in the pattern as long as we know its position.

Reflect

Estimate how long it would take to find the 500th term in Worked example 6.6 by starting with 4 and adding 4 each time. Assume you could do each addition in ten seconds.

How does this compare to the time it takes to multiply 4 by 500?

This new approach is useful for practical applications of patterns.

6.3.2 Variables and rules

Explore 6.5

Number patterns occur whenever two quantities are related to one another by a fixed rule. There are many examples from science, economics, statistics and other fields. For example, the time it takes to travel a certain distance is related to the speed at which you are travelling.

What examples can you think of from your own experience?

Symbolic or **algebraic notation** allows us to perform calculations more quickly and easily. Instead of using wordy descriptions of quantities, we represent them with a single letter, usually taken from the Latin or Greek alphabets.

Hint

It is a good idea to choose letters based on what they stand for: c for cost, n for number, t for time, and so on.

Any letter we use to represent the value of a quantity is called a **variable**. For example, we could choose the letter n to represent the number of gigabytes used by a student, and the letter c to represent the cost paid to the mobile-phone company.

We can use variables to translate rules into mathematical **statements** or **expressions**. A rule that is particularly important or used often is called a **formula**. We can use mathematics to manipulate formulas so that we can calculate the values we need.

The use of variables, expressions and formulas to express and analyse situations mathematically is called algebra.

The formula relating time, distance and speed says that the slower you are travelling, the longer it will take you to reach your destination.

Worked example 6.7

a   Express the number pattern 4, 8, 12, ... in algebraic notation.

b   Calculate the 200th term in this pattern.

c   Find the position of the term in the pattern that equals 2020.

Solution

a   We saw in Worked example 6.6 that the rule for this pattern is 'the value of each term in the pattern is 4 multiplied by the term's position'.

To write this in algebraic notation, we need to choose variables to represent the value and the position of each number.

We can use the variable v to stand for the term *value*, and the variable p to stand for the term *position*. Then we can rewrite the table from Worked example 6.6 using these variables.

Term position, p	1	2	3	4	5	...
Term value, v	4	8	12	16	20	...

Now we can restate the rule 'The value of each term in the pattern is 4 multiplied by the term's position.' as:

$v = 4 \times p$

We can use this to answer parts b and c.

b p stands for the term position, so for the 200th term $p = 200$

The value of the 200th number in the pattern is the value of v when p is replaced with 200:

$v = 4 \times p$

$v = 4 \times 200 = 800$

So, the value of the 200th term is 800.

c This time we know the value of the term and we need to find its position.

As before, we can replace the variable in the formula with the information we know. In this case, we replace v with 2020.

$v = 4 \times p$

$2020 = 4 \times p$

Now we have to find the value of p for which $4 \times p$ equals 2020. This is called **solving** the equation $4 \times p = 2020$ for the unknown value of p.

If $4 \times p = 2020$ then p must be equal to 2020 divided by 4.

$p = \dfrac{2020}{4} = 505$

So, 2020 is the 505th term.

 Fact

When many expressions are equal to each other, we can use a chain of equalities. If $a = b$ and $b = c$ we can use the transitive property of equality and write $a = c$

Thus, $v = 4 \times 200 = 800$ means $v = 800$

 Connections

You will find out more about solving equations in Chapter 7.

Being able to write and solve questions using algebra is useful in many practical situations. In the next worked example, you will see how it can be used to answer the question from Worked example 6.5 in a more efficient way.

Worked example 6.8

A mobile-phone company charges its customers $3.00 a month for a contract. In addition to this fixed cost, each gigabyte of data traffic used costs $0.50.

a In October, Julia used 6 gigabytes. How much did she spend in October?

b In November, Julia was billed $7.50. How many gigabytes did she use?

Solution

We saw in Worked example 6.5 that the rule for this problem is 'Start with $3.00 and add $0.50 for every gigabyte.'

We can use the variable n to represent the number of gigabytes Julia uses and the variable c to represent the cost in dollars of her monthly subscription. This is a table of the first few values.

Number of gigabytes used, n	0	1	2	3
Cost ($), c	3.00	3.50	4.00	4.50

When 0 gigabytes are used, the cost is $3.00 plus $0 \times \$0.50$

When 1 gigabyte is used, the cost is $3.00 plus $1 \times \$0.50$

When 2 gigabytes are used, the cost is $3.00 plus $2 \times \$0.50$

When 3 gigabytes are used, the cost is $3.00 plus $3 \times \$0.50$

So, the rule is:

$c = 3.00 + 0.50 \times n$

Now we can use the formula to answer the questions.

a   We need to find the cost when the number of gigabytes is 6, so we replace n in the formula with 6.

$c = 3.00 + 0.50 \times n = 3.00 + 0.50 \times 6$

$= 3.00 + 3.00 = 6.00$

So, by using the transitive property of equality, the cost for 6 gigabytes is $c = \$6.00$

b   We need to find the number of gigabytes when the cost is $7.50, so we replace c in the formula with 7.50.

$7.50 = 3.00 + 0.50 \times n$

Now we need to find the value for n in this equation.

Subtract 3.00 from both sides. This means that $4.50 = 0.50 \times n$

To find n, we divide both sides of this equation by 0.50.

This means that $n = \dfrac{4.50}{0.50} = 9$

So, Julia used 9 gigabytes in November.

Looking back, we see that both answers are the same as the answers we found using the table in Worked example 6.5.

Fact

We usually leave the units ($ in this case) out of the formula. We must remember to include the units in the table heading and the final answer.

Look back at the methods used in Worked examples 6.5 and 6.8.
Which method do you prefer? Why do you prefer it?

Investigation 6.2

1 A city car uses 5 litres of fuel every 100 kilometres.

What rule relates the distance travelled, d (in kilometres), to the amount of fuel used, f (in litres)? You can use a table of values to help you work out the rule.

What distance does the car travel if it uses 7 litres of fuel?

How many litres of fuel are required to travel 80 kilometres?

2 The population of elephants in a wildlife park grows by 10% every year. There were 200 elephants the year the park was created.

How many elephants will be in the park after 1 year? How many will there be after 2 years?

What rule relates the number of elephants in the park, E, to the number of years, n, since the park was created?

3 Look back at the definition of arithmetic and geometric patterns in Section 6.2.

Describe how you can check if a rule describes an arithmetic or geometric pattern.

What type of pattern does the rule in question 1 describe?

What type of pattern does the rule in question 2 describe?

Reflect

When we model a real-world process with a rule, we often have to make some **assumptions** about the process.

What assumptions were made in modelling the distance travelled by the car?

What assumptions were made in modelling the population growth for the elephants?

Often the variables we use in mathematics do not represent a particular object in the real world. A variable can stand for any number and we need to be able to work with them even if we do not know what they represent.

Worked example 6.9

a   Evaluate $y = 2 + x$ when $x = -3$

b   Complete the table for the rule $y = 2 + x$

x	1	2	3	4
y				

Solution

a   To evaluate an expression, we need to replace a variable with a known value, then calculate the remaining unknown value.

In this case, we need to replace x with -3.

$y = 2 + x$

$y = 2 + (-3)$

$y = -1$

b   To fill in a table like this, we need to evaluate y for each value of x given in the table.

For $x = 1$, we have $y = 2 + 1 = 3$

For $x = 2$, we have $y = 2 + 2 = 4$

We can continue in the same way to complete the table.

x	1	2	3	4
y	3	4	5	6

 Research skills

 Investigation 6.3

Do you think that algebra will work with recursive rules? In other words, can a rule such as 'Add 3 to the previous number.' be translated into an algebraic expression?

Ask your teacher to help you research 'recursively defined sequences'.

Asking your teacher for guidance when you research a topic not covered in the textbook helps you find the most relevant sources for you. Search engines do not know what maths you already know!

 Practice questions 6.3

1   For each rule below, evaluate y when $x = 2$

   a $y = x - 2$ **b** $y = 2 \times x - 2$

   c $y = 3 \times x + 10$ **d** $y = 3 - x$

e $y = 1 - x$

f $y = -x$

g $y = \dfrac{12}{x}$

h $y = x \times x$

i $y = 3 \times x \times x$

j $y = \dfrac{12}{x \times x}$

k $y = (3 + x) \times 3$

l $y = (3 + x) \times x$

2   For each rule below, evaluate p when $q = -1$

a $p = 2 \times q$

b $p = q + 4$

c $p = q - 1$

d $p = -1 - q$

e $p = 3 \times q + 3$

f $p = 3 \times (q + 1)$

g $p = \dfrac{-1}{q}$

h $p = q \times q$

i $p = q \times (q + 2)$

j $p = q \times (q + 1)$

k $p = (q + 2) \times (q + 3)$

3   Rewrite each rule in algebraic notation. Choose appropriate variables for each question.

Example: area of a rectangle = base times height

Answer: $A = b \times h$

a   area of a square = side times side

b   total cost = unit cost times number of items

c   perimeter = sum of the lengths of the three sides

d   number of students = sum of students in class A, class B and class C

e   speed = distance divided by time

f   average of two numbers = sum of the two numbers, divided by two

g   Frank is three years older than his sister Louise.

4   Copy and complete each table using the given rule.

a $w = s + 2$

s	1	2	3	4
w				

b $q = 2 \times r$

r	0	2	4	6
q				

c $y = 10 - x$

x	1	2	3	4
y				

 Fact

You could write part g as 'Frank's age is equal to 3 plus Louise's age' before using variables.

d $l = \dfrac{t}{2}$

t	1	2	3	4
l				

e $w = 2 \times s + 1$

s	1	2	3	4
w				

f $y = 10 - 2 \times x$

x	0	1	2	3
y				

g $y = (4 - x) \times (x + 1)$

x	1	2	3	4
y				

h $y = \dfrac{24}{x}$

x	1	2	3	4
y				

i $y = x \times x$

x	0	1	2	3
y				

 **Challenge Q5**

5 For each table of values, write the rule using the given variables.

a
x	0	1	2	3
y	5	6	7	8

b
a	2	4	6	8
b	2	4	6	8

c
t	1	2	3	4
q	2	1	0	−1

d
w	2	4	6	8
d	10	20	30	40

e
a	2	4	6	8
b	1	2	3	4

f
x	−1	0	1	2
y	−5	0	5	10

g
a	−1	0	1	2
b	−6	−1	4	9

h
a	1	2	3	4
b	3	6	9	12

i	x	1	2	3	4
	y	12	6	4	3

j	a	−5	0	5	10
	b	5	0	−5	−10

6 Entry to a swimming pool costs €3.00. Every hour spent at the pool costs an additional €0.50.

    a Write down the rule that relates the number of hours at the pool, h, and the cost, c.

    b On a hot summer day, Andrea spends 8 hours at the pool. How much does this cost?

    c Francesco spends part of a day at the pool. He is charged €6.50. How much time did he spend at the pool?

7 Gardens of different size are surrounded by square paving tiles.

 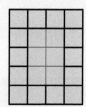 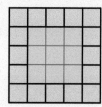

    a Copy and complete the table of values for the size of the garden, s, and the number of tiles around it, n.

s	3			
n	12			

    b Write down the rule that relates the size of the garden, s, to the number of tiles around it, n.

    c How many tiles would surround a garden of size $s = 18$?

    d A garden is surrounded by 20 tiles. What is the size of the garden?

8 A skyscraper has 20 floors. The elevator takes 30 seconds to travel from the ground floor to the top floor.

    a Write down the rule that relates the floor number, n, to the time taken, t, to reach it from the ground floor.

    b How many tiles would surround a garden of size $s = 18$?

    c A garden is surrounded by 20 tiles. What is the size of the garden?

Reflect

Explain how you came up with your rule in question 7b. Try to explain your reasoning using accurate terminology.

In question 8a, what assumption did you make about the motion of the elevator?

6.4 Graphing rules

In this section, we will investigate one of the most useful tools in mathematics: graphs. Graphs are excellent tools for helping us to visualise patterns.

You are already familiar with the number line: we can draw any real number as a point on a number line. The number line must have a starting point, which we call the **origin**, and a **scale**, which defines how far apart the numbers are.

Now that we know about variables, we can label the number line with the letter n for number.

 Explore 6.6

Consider the pattern below. Let n be the number of circles used. We can arrange the pattern in a table. n represents the position and C represents the number of circles.

n	1	2	3	4	...
C					...

Here is a number line. What does the red dot on this number line represent?

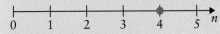

Look at the pattern. It relates values of n to values of C. For example, the pattern relates $n = 3$ to $C = 9$. The pair $(3, 9)$ belongs to the rule.

If we want to represent the rule with a graph, we need a way to plot a point that represents a pair of numbers.

We could use another number line to represent 9.

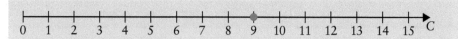

How could you distinguish between the number lines for C and for n to show both parts of the pair?

n	2	3	4	5
p	4	5	6	7

Let us look at how to represent the pair $(4, 6)$ on a graph. One possibility is to draw the p line **perpendicular** to the n line. You need to make sure that the origins (the zeros) of both lines are in the same place.

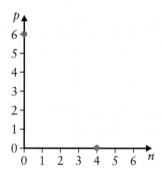

This system of two number lines is called a **number plane** or a **Cartesian plane**. Each number line is called an **axis** of the Cartesian plane. The plural of axis is **axes**. We can then plot the pair $(4, 6)$ like this.

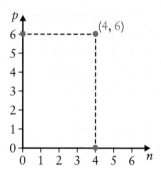

Each **pair** of numbers in a pattern becomes a **point** on the Cartesian plane. We say that the point $(4, 6)$ has $n = 4$ and $p = 6$. We call the pair of values written in brackets, $(4, 6)$, the **coordinates** of the point.

Numbers in the *top row* of the table of values go on the *horizontal* axis and are written *first* in the brackets.

Numbers in the *bottom row* go on the *vertical* axis and are written *second* in the brackets.

 Fact

The Cartesian plane is named after René Descartes, a French mathematician and philosopher of the seventeenth century. According to the story, he devised the grid idea while watching a fly walk across the ceiling in his room as he lay in bed.

 Connections

The horizontal line is also called the abscissa axis, and the vertical line is also called the ordinate axis. You will use this again in Chapter 7.

Worked example 6.10

Plot all points in the table of values on the Cartesian plane.

n	2	3	4	5
p	4	5	6	7

Solution

We start by drawing and labelling the horizontal and vertical axes. The variable in the top row is n, so we label the horizontal axis n. The variable in the bottom row is p, so we label the vertical axis p.

We need to choose a sensible scale. The values in the table range from 2 to 7. We need to fit all these values on the graph.

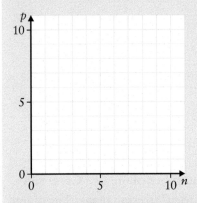

The first column in the table has $n = 2$ and $p = 4$. The point will have coordinates $(2, 4)$. Imagine a vertical line going through 2 on the n axis, and a horizontal line going through 4 on the p axis. We draw the point where these two lines cross. We can plot the other points in the same way.

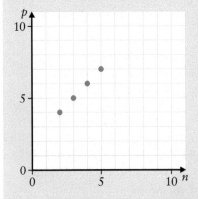

Communication skills

Neat and tidy graphs communicate your findings in a clearer way. There are many graphing programs available that you can use to produce high-quality graphs. Check with your teacher or librarian to find out what resources you can use.

Look at the graph in Worked example 6.10. What can we observe? Would we be able to add other points on the graph that follow the same trend? Could we use the extra points to add more columns to the table of values?

Does graphing a table of values help us spot the pattern? Write down the rule using the variables n and p.

How many numbers are needed to plot **one** point on the number **line**?

How many numbers are needed to plot **one** point on the Cartesian **plane**?

What can a triple, such as (4, 6, 11), stand for? How can we represent it as point?

 Fact

The link between a pair of numbers and one point on a plane lays the foundations of a branch of mathematics called analytical geometry.

We have seen that a pattern can be given as a table of values:

Number of gigabytes used, n	0	1	2	3	4	5	6	7	8	9
Cost ($), c	3.00	3.50	4.00	4.50	5.00	5.50	6.00	6.50	7.00	7.50

 Connections

Graphs are used in many other fields to make it easy for the reader to understand the information. Where else have you seen graphs, both at school and outside?

We have also seen a pattern given as an algebraic relation:

$c = 3.00 + 0.50 \times n$

Now we can see the same pattern given as a set of points on a graph.

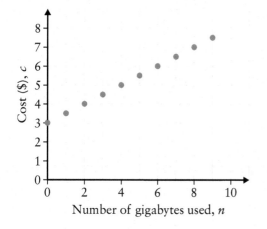

 Hint

Your calculator or computer can plot such patterns too. Learn how your specific calculator does this. Here is a sample.

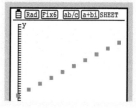

Think about the advantages and disadvantages of the table of values, the algebraic relation, and the graph. Which of them is best? Will it be best in all circumstances?

Practice questions 6.4

1   For each graph, copy and complete the table of values.

a

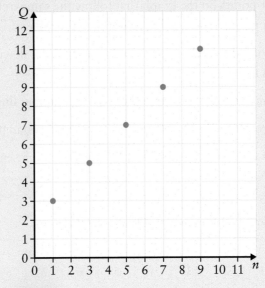

n	1	3	5	7	9
Q					

b

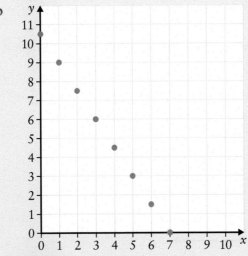

x	0	1	2	3	4	5	6	7
y								

c

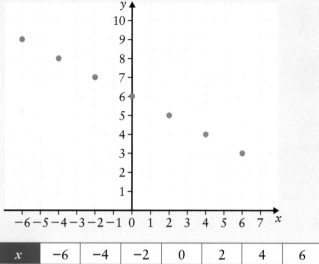

x	−6	−4	−2	0	2	4	6
y							

d

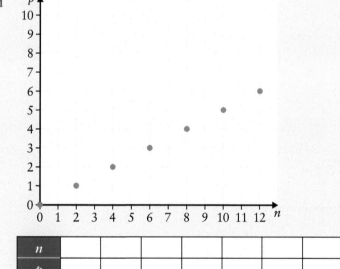

n							
p							

2   For each table of values, draw a Cartesian plane, graph the pattern and label the axes.

 Hint Q2

We can use a calculator or computer to graph each pattern.

a

x	1	2	3	4	5	6	7
y	2	4	6	8	10	12	14

b

x	1	2	3	4
y	12	6	4	3

c

x	1	2	3	4	5	6	7
y	12	10	8	6	4	2	0

d

x	−3	−2	−1	0	1	2	3
y	12	10	8	6	4	2	0

e

x	−3	−2	−1	0	1	2	3
y	−2	−1	0	1	2	3	4

f

x	−3	−2	−1	0	1	2	3
y	5	2	−1	−4	−7	−10	−13

3   For each pattern rule, complete a table of values for $x = 0, 1, 2, 3, 4, 5$
    Then graph the pattern on a Cartesian plane.

   a $y = 2x$ b $y = 3x + 1$

   c $y = 5 - x$ d $y = \dfrac{1}{x + 1}$

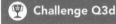
Challenge Q3d

4   Write down an algebraic expression for the rule in each of the graphs
    below.

Hint Q4a

We can use a calculator
or computer to verify
expressions. Here is an
example for part a.

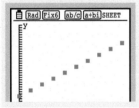

a

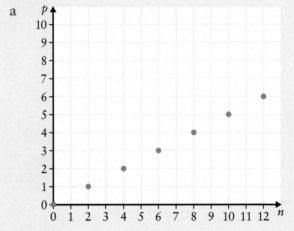

b

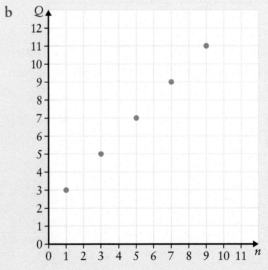

c

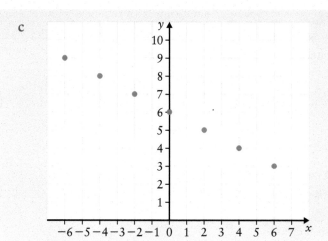

d

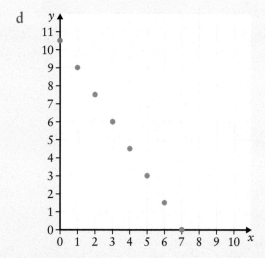

5   Graph the rules in questions 6, 7 and 8 in Practice questions 6.3.

? **Check your knowledge questions**

1 a Draw the next two figures in the pattern.

  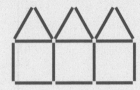

b Copy and complete the table with the missing numbers.

Position in pattern	1	2	3	4
Number of sticks	6			

c State the rule in words.

d Choose suitable variables and write an algebraic expression for the rule between the position in the pattern and the number of sticks.

2 Write down the first four numbers for each rule.

a Start with 2, add 4.

b Start with 5, subtract 10.

c Start with 2, multiply by 2.

d Start with 3, multiply by −2.

e Start with 2, add 1 then multiply by 3.

f Start with 2, multiply by 3 then add 1.

3 Complete each pattern with the missing numbers.

a ☐, 3, 6, 9, ☐   b 3, 5, 7, ☐, ☐

c ☐, ☐, 4, 16, 64   d ☐, 2, ☐, 6, 8

e ☐, 8, 4, 0, ☐   f −3, −5, ☐, ☐, −11

g 4, −4, ☐, ☐, 4   h 2, 4, ☐, ☐, 10

i 2, 4, ☐, ☐, 32   j 2, 4, ☐, ☐, 65 536

4 Evaluate y for the given value of x.

a $y = 7x − 2$ for $x = 2$ b $y = 2x − 7$ for $x = 3$

c $y = 7 × (x − 2)$ for $x = 2$ d $y = \dfrac{7}{x − 2}$ for $x = 9$

e $y = 2x − 4$ for $x = −1$ f $y = 4 + x$ for $x = −1$

g $y = 4 − x$ for $x = −1$ h $y = −4 × (−4 − x)$ for $x = 1$

i $y = −4 × (−4 − x)$ for $x = −1$

5　Copy and complete each table using the given rule.

a　$y = x + 2$

x	1	2	3	4
y				

b　$y = -x + 2$

x	1	2	3	4
y				

c　$y = 2x - 2$

x	−3	−2	−1	0
y				

d　$y = 2 - x$

x	1	2	3	4
y				

e　$y = 2 - x$

x	0	0.5	1	1.5
y				

f　$y = x + 2$

x	−1.5	−1	−0.5	0
y				

g　$y = -x + 1$

x	−1.5	−1	−0.5	0
y				

6　For each table of values, write the rule and graph it.

a

x	0	1	2	3
y	4	5	6	7

b

x	−2	0	2	4
y	1	3	5	7

c

x	−4	−3	−2	−1
y	−1	−3	−5	−7

d

x	−10	−5	0	5
y	7	4	1	−2

7　For each graph, build a table of values and write the rule.

a

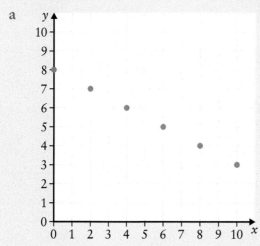

213

b
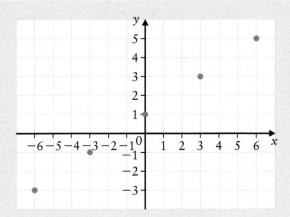

c
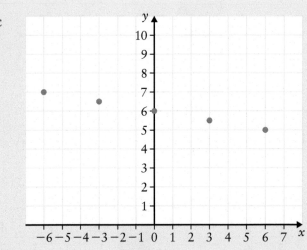

d
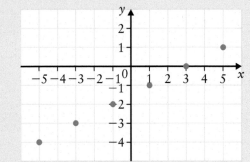

🛡 Hint Q8d

What happens when we try to divide by 0?

The table of values will have one missing number, and one point will be missing from the graph.

8   For each rule, build a table of values and draw the graph for the x values $-4, -2, 0, 2$ and 4.

a $y = 2x + 1$ b $y = 3x - 2$

c $y = (x - 3) \times (x + 3)$ d $y = \dfrac{4}{x}$

Algebra and equations

7 Algebra and equations

KEY CONCEPT

Form

RELATED CONCEPTS

Equivalence, Representation, Simplification

GLOBAL CONTEXT

Scientific and technical innovation

Statement of inquiry

Generalising patterns and representing them in a simplified form helps us to find innovative solutions to real-life problems.

Inquiry questions

Conceptual

- Can mathematics be regarded as a language?

Factual

- How do mathematical skills support technical advancement?

Debateable

- Can all problems be tackled efficiently with mathematical tools?

1   What is a pattern? Name three ways of describing a pattern.

2   What advantages do algebraic rules have over the other methods
     to describe a pattern?

3   a   How are pairs of numbers placed on the Cartesian plane?

     b   Copy the set of axes below, and plot the points (−1,2), (3,3),
          (−2,−2) and (0.5, −1.5).

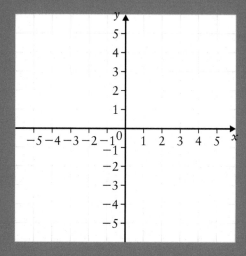

4   a   How do we graph a rule?

     b   Graph the rule $y = 2x + 2$

7 Algebra and equations

 7.1 **Algebraic notation**

 Explore 7.1

Look at the expression $y = x + x + x + x$

Copy and complete the table of values for this expression.

x	1	2	3	4
y				

By looking at the table of values, can we write down another rule for this pattern? Which of the two algebraic rules is simpler?

This expression is more complicated: $y = \dfrac{(x + 2) \times (x - 2)}{x \times x - 4}$

Copy and complete the table of values for this expression.

x	5	6	7	8
y				

By looking at the table of values, can you write down another rule for this pattern? Which of the two algebraic rules is simpler?

In Explore 7.1, we saw that algebraic expressions sometimes look more complicated than they really are. We found a way to make them easier to work with.

In this chapter, we will learn how to make an algebraic expression simpler without using a table of values.

7.1.1 Terminology

A variable is a letter or symbol used to represent a quantity whose value can vary.

In an expression that is a sum, such as $x + 34 + b$, the quantities that are being added are called **terms**. The terms of the expression $x + 34 + b$ are x, 34 and b.

In an expression that is a product, such as $2 \times a \times n$, the quantities that are being multiplied are called **factors**. The factors of the expression $2 \times a \times n$ are 2, a and n.

Fact

When numbers and variables are combined using the four operations $+$, $-$, $\times$, $\div$ and powers, we obtain an algebraic expression.

A labyrinth looks like a complicated structure, but there is only one path to follow. If you keep walking, you will get to the centre. Sometimes, algebraic expressions also look complicated, but if you work through them step by step, you can make them simpler.

Connections

You learned about using variables to represent numbers in Chapter 6.

Reminder

The words term and factor are not synonyms! Use them accurately.

7.1.2 Abbreviations

When we write multiplications in algebra, we leave out the × sign when possible:

- in the product of a number and a variable: $12 \times n = 12n$
- in the product of two variables: $a \times b = ab$
- with brackets: $3 \times (x + 1) = 3(x + 1)$ and $a \times (b + c) = a(b + c)$

When we multiply two numbers we need to keep the multiplication sign:

- 3×7 cannot be written as 37

Often the dot symbol $\cdot$ is used for multiplication:

- $3 \cdot 7 = 3 \times 7$

When we multiply a variable by the number 1 or −1, we do not write the number 1:

- $1 \times n = n$
- $-1 \times n = -n$

When we multiply a variable by the number 0, we do not write the variable:

- $0 \times n = 0$

For division, we usually use the fraction sign instead of the ÷ sign:

- $3 \div y = \dfrac{3}{y}$
- $a \div b = \dfrac{a}{b}$

If we divide a variable by a number, we have two ways of writing the same expression:

- $x \div 3 = \dfrac{x}{3}$
- $x \div 3 = \dfrac{1}{3}x$

When we divide a variable by the number 1, the fraction is unnecessary:

- $\dfrac{n}{1} = n$

For repeated multiplication, or powers, we use the same notation as with numbers:

- $x \times x = x^2$
- $x \times x \times x = x^3$

In the power x^n, x is called the **base** and n is called the **exponent** or **index**. The plural of index is **indices**.

 Fact

In the product of a number and a variable, we always write the number first.

Both $12 \times n$ and $n \times 12$ are written as $12n$.

 Hint

Distinguish between $3 \cdot 7$ and the decimal 3.7.

 Fact

Out loud, $a \div b = \dfrac{a}{b}$ is often read as 'a over b'.

 Fact

Out loud, x^2 is read as 'x squared' and x^3 is read as 'x cubed'.

Explore 7.2

If you hear someone say 'three times four plus five', how would you write this down? Is this the only way?

Can you come up with other expressions that would sound the same when read out loud but could be written in different ways?

When we read expressions with brackets out loud, we must make sure we are very clear. One way to read $a(b + c)$ is 'a, open bracket, b plus c, close bracket'. Can you think of any other ways that make the expression clear?

Why do we use brackets? How are brackets used in everyday writing? Is this different to how they are used in mathematics?

Worked example 7.1

Rewrite each expression without using multiplication or division signs.

a $7 \times b$ **b** $3 \cdot x$ **c** $3 \div a$ **d** $(3 + x) \div y$

Solution

a To write an algebraic product without using a multiplication sign, we leave out the sign and write the factors next to each other.

$7 \times b = 7b$

b $3 \cdot x = 3x$

c To write a division without using a division sign, we write it as a fraction. The number or variable to the left of the division sign is the numerator and the number or variable to the right of the division sign is the denominator.

$3 \div a = \dfrac{3}{a}$

d If there is an expression in brackets, the entire expression becomes the numerator or denominator.

$(3 + x) \div y = \dfrac{3 + x}{y}$

Learning vocabulary and grammar is vital to be able to communicate effectively, both with spoken languages and the language of mathematics.

Investigation 7.1

In any language, there are conventions that we must be familiar with. Can you think of some of these conventions:

*   in your native language (if it is not English)
*   in the English language
*   in the language of mathematics?

You could start with the apostrophe symbol '. In how many different ways have you seen this symbol used?

7.1.3 Grouping symbols

In Chapter 1, you learned about the order of operations and about using brackets when you need to override the normal order of operations.

There are other grouping symbols you can use to make your expressions clear. Two that you are likely to see are:

- the fraction bar —
- the square root sign $\sqrt{}$

The expressions $\dfrac{4+7}{3}$ and $\dfrac{1}{x+1}$ are interpreted as $\dfrac{(4+7)}{3}$ and $\dfrac{1}{(x+1)}$

The expression $\sqrt{4+9}$ means $\sqrt{(4+9)}$

 Reminder

The order of operations is: brackets, powers, then multiplications and divisions, then additions and subtractions.

 Reflect

'Algebraic notation is more precise than any language.'

Do you agree with this statement?

 Thinking skills

 Communication skills

 Investigation 7.2

Look at the expression $3 \div 2x$

Some people would say that because the 2 and the x have no symbol between them, they should be treated as a single entity, so the expression could be written as $\dfrac{3}{2x}$

Other people would say that because division and multiplication have the same priority, they should be carried out from left to right, so the expression could be written as $\dfrac{3}{2} \times x$ or $x\left(\dfrac{3}{2}\right)$

Write some expressions that could be interpreted in more than one way. How could you rewrite your expressions so that the meaning is clear?

 Reminder

The rules you have learned for numbers apply in exactly the same way to expressions containing variables. Since variables stand for numbers, they obey the same rules as numbers.

 Practice questions 7.1

1 Write each expression without using a multiplication sign.

  a $3 \times g$ b $n \times 2$ c $3 \cdot y$ d $14 \cdot b$

  e $x \cdot (-1)$ f $-3 \times g$ g $1 \cdot y$ h $2 \cdot (y + z)$

  i $\dfrac{1}{2} \times w$ j $3 \times a + 4 \times b$ k $4 \cdot a - 5 \cdot b$ l $1 \times a + 10$

2 Write each expression as a fraction.

  a $3 \div g$ b $n \div 2$ c $x \div 3$ d $15 \div b$

  e $2b \div 3$ f $x \div q$ g $x \div 1$ h $1 \div x$

  i $2 \div (y + z)$ j $(x + 1) \div (x - 1)$ k $3 \div (2x)$

 Fact Q2g and Q2h

Division is not commutative. This means that the result of a division changes if the order of the numbers is reversed. For example, $6 \div 3 = 2$ but $3 \div 6 = 0.5$

Connections

In calculators, the order of operations is built in by the manufacturer.

Try evaluating $\dfrac{12}{2 \times 1.5}$ by entering $12 \div 2 \times 1.5$

What result do you get? What is the correct result? If your calculator gave a different result, why do you think that happened?

The calculator interprets $12 \div 2 \times 1.5$ as $(12 \div 2) \times 1.5$, but $\dfrac{12}{2 \times 1.5}$ means $12 \div (2 \times 1.5)$

When in doubt, always use brackets. Most calculators also have a fraction template.

3 Starting from the numbers 2 and 3, we can make larger numbers by using different combinations of operations and brackets. For instance, we could make:

$2 + 3 = 5$

$2 \times 3 = 6$

$2^3 = 2 \times 2 \times 2 = 8$

$3^2 = 3 \times 3 = 9$

The largest possible number is 9.

Find the largest number that can be made with each group of numbers.

a   4 and 5   b   1 and 3   c   2, 3 and 4   d   1, 3 and 5.

4 Rewrite each expression, showing all multiplication and division signs and all brackets. For example, $3x + 5 = 3 \times x + 5$, $\dfrac{1 + x}{3} = (1 + x) \div 3$

a $2s + t$ b $2(s + t)$ c $3(x - y)$ d $3x - y$

e $3xy$ f $\dfrac{3x}{y}$ g $2x - 4y$ h $2(x - 4y)$

i $2 + \dfrac{x}{3}$ j $\dfrac{2 + x}{3}$ k $x - \dfrac{4 + s}{w}$ l $x - 4 + \dfrac{s}{w}$

m $x^2 + 1$ n $(x + 1)^2$ o $a^2 - b^2$ p $(a + b)(a - b)$

q $\dfrac{q + b}{q - b}$ r $\dfrac{q + b}{q} - b$ s $\dfrac{1}{2s}$ t $\dfrac{1}{2}s$

u $\dfrac{s}{2}$ v $\dfrac{3(x + 1)}{y}$ w $\dfrac{3x}{2y}$ x $\dfrac{3x}{2}y$

y $\dfrac{3}{2}xy$ z $\dfrac{3}{2xy}$

5 Explain the meaning of each expression in question 4, using 'first' and 'then' to clarify the order of operations. For example, $3x + 5$ means 'first multiply 3 by x then add 5' and $\dfrac{1 + x}{5}$ means 'first add 1 and x then divide the result by 5'.

 7.2   Algebraic expressions

In this section, we will learn how to evaluate, simplify and expand algebraic expressions.

7.2.1 The meaning of algebra

 Explore 7.3

When tables are set at a restaurant, two plates are arranged for each customer. There is also one plate of bread for each table.

If two customers sit at a table, how many plates will be arranged on their table? Explain how you know.

After dinner, the waiter collects 13 plates from another table. How many customers were seated at that table? Explain how you know.

 Connections

Remember Pólya's four-step problem-solving process from Chapter 2 as you tackle this question: understand the problem, make a plan, carry out the plan, look back.

When you worked through the problems in Explore 7.3, you might have thought of using algebra as part of your plan. Algebra can help us answer questions like this in an efficient way.

 Worked example 7.2

A box of pencils has been delivered to a classroom.

The pencils have been packaged in two different ways:

- There are seven loose pencils.

- There are two identical bags, each containing the same unknown number of pencils.

The label on the box says, 'weight of one pencil = 10 grams, total shipping weight = 1570 grams'.

Write an equation to represent this information.

Solution

To write an equation to represent the information, we need a way to represent the number of pencils in each bag. This number could be anything, say 12, 57 or 132. We can represent this number with a variable, x.

x = number of pencils in each bag

The information we know is:

- We have 7 loose pencils.

 Reminder

We could use any letter to represent the number of pencils, for example, p for pencil or u for unknown. x is a common letter to use.

- We have 2 bags, each containing x pencils.
- One pencil weighs 10 grams.
- All the pencils together weigh 1570 grams.

All the pencils are identical, so we can relate the number of pencils to their weight. If one pencil weighs 10 grams, then two pencils will weigh $2 \times 10 = 20$ grams, and x pencils will weigh $x \times 10 = 10x$ grams. So:

Weight of pencils in each bag = $10x$

There are two bags, so:

Weight of pencils in both bags = $10x + 10x$

There are also seven loose pencils, so:

Weight of loose pencils = 7×10

The total weight of all the pencils is the sum of the weights of the pencils in the two bags and of the loose pencils:

Total weight of pencils = $10x + 10x + 7 \times 10$

We now have an algebraic expression for the total weight of pencils in the box. There is one piece of information that we have not used yet: the total weight of all the pencils is 1570 grams. We can equate this expression to the known weight to make an **equation**.

$10x + 10x + 7 \times 10 = 1570$

 Reminder

When we build algebraic expressions we often drop the units.

 Fact

Later in the chapter, we will **solve** the equation. This means finding the value of x that makes $10x + 10x + 7 \times 10$ evaluate to 1570.

 Reflect

What are the advantages of using x to represent a unknown quantity?

 Practice questions 7.2.1

1   Write an expression for the total number of pencils using variables.

a

b

c

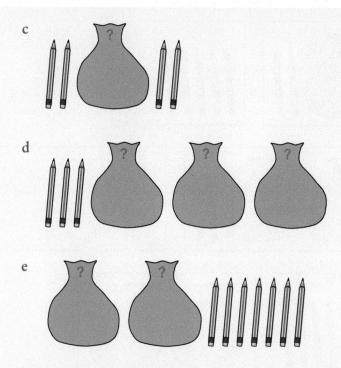

d

e

2   For each set of two boxes, write an expression for the total number of
    pencils using variables and brackets. For example:

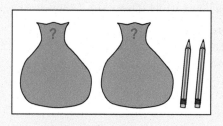   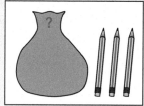

$(x + x + 2) + (x + 3)$

a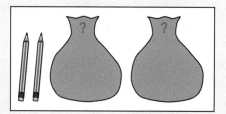

b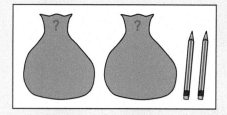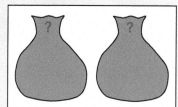

c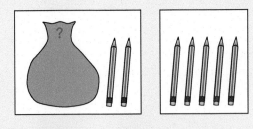

d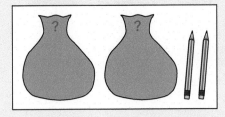

e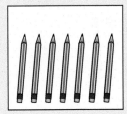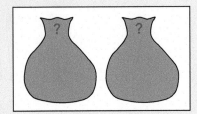

3   Each expression represents the contents of a box of pencils, with x representing the number of pencils in a bag. Draw the contents of each box based on the algebraic expression.

a $x + 2$

b $3 + 4x$

c $(3) + (x + 1)$

d $(2 + 2x) + (x)$

e $(2 + x) + (2x + 1)$

7.2.2 Simplifying expressions

Explore 7.4

Imagine you have 2 apples in your left hand and 3 apples in your right hand.

How many fruits do you have in total?

Can you make the expression 2 + 3 simpler?

Now imagine you have 2 apples in your left hand, another 3 apples in your right hand, and 4 oranges in a bag.

How many fruits do you have in total?

Can you make the expression 2 🍎 + 3 🍎 + 4 🍊 simpler?

Algebraic expressions are sometimes complicated. We can often write them in a simpler, shorter way by performing some operations on the variables.

We know that repeated addition is made shorter by using a multiplication sign. $3 + 3 + 3 + 3 + 3$ can be written as 3×5 or $3 \cdot 5$. In the same way, we can write $x + x + x + x + x$ as $x \times 5$, $x \cdot 5$, or $5x$.

When two terms have the same combination of letters or symbols, we call them **like terms**.

For example, 2 🍎 and 3 🍎 are like terms, because they both have the

same symbol. We have 5 🍎 in total.

On the other hand, 3 🍎 and 4 🍊 are *not* like terms, because the

symbols are different.

Look back at the expression from Worked example 7.2: $10x + 10x + 7 \times 10$ Can we simplify this expression?

The expression $10x + 10x + 7 \times 10$ has three terms ($10x$, another $10x$ and 7×10).

We can immediately simplify 7×10 to 70.

$10x$ and $10x$ are like terms, so we can simplify these by adding the 10s: $10x + 10x = 20x$

The simplified expression is now: $20x + 70$

$10x$ and 70 are not like terms because they do not have the same symbol. The expression cannot be simplified any further.

💬 **Reminder**

Term is the name we give to each quantity that is being added in a sum.

◈ **Fact**

Adding or subtracting like terms is called **collecting like terms**.

Practice questions 7.2.2

1 Write down the number of terms in each expression.

 a $7x + 2$ b $x + 2y + 1$ c $a - b + c$

 d $w + 2q$ e $1 - 2x + 3y - 4z$ f $s + 2t - q$

2 Find the like terms in each group.

 a $2x, 3, 4x$ b $1, 3w, 4$

 c $w, r^2, -4w$ d $2x^2, 4y, -3x^2, y$

 e $1, -2s, 4, t$ f $xy, 2x, -3xy, y$

 g $2mn, 3n, 4m, -mn, 1$ h $9n^2, 3m, -5mn, 3n^2, 2mn$

 i xy, yx, x^2, y, xy^2 j $2st^2, 3t^2, -4, -t^2, st^2, 1$

 k $xyz, xy^2, yz^2, -2xyz, -2yz^2$

3 Simplify each expression by collecting like terms.

 a $2x + 3x$ b $8y - 4y$ c $a + 2a$

 d $3t + 4t$ e $5w - 3w$ f $x + 3x - 2x$

 g $x + 3x - 4x$ h $w - 3w + 5w$ i $xy + 2xy$

 j $15x - 14x$ k $4ab + ba - 2ab$ l $2x + 3 - x + 4$

4 Simplify each expression by collecting like terms.

 a $2x + 3x + y + 2y$ b $-3w + 2t - w + 3t$

 c $x + 3 + 3x - 10$ d $2x - x + y$

 e $x^2 + 2x^2 + y + 2y$ f $x^2 + 2x^2 + x + 2x$

 g $6x + 3 - 4x$ h $4 + t + 5t - 6$

 i $10a^2 - 4 - 3a^2 + 5$ j $2x^2 + 3t - x^2 + t$

 k $10j + i + i^2 - 5j$ l $-a + b - 3b + a$

 m $7a^2 + a - 3a - 4a^2$ n $4ab + a - 3ab + b$

 o $x + xy - x^2 + 2xy - x + 2x^2$

5 a Alice buys 3 boxes of chocolates, while Bob buys 5. If there are x
      chocolates in each box, how many chocolates do Alice and Bob
      have altogether?

 b One kilogram of apples contains n apples. Aishwarya buys 3 kg of
    apples, but her bag splits and 1 kg is lost. How many apples does
    she have now?

Challenge Q5b

7.2.3 Evaluating expressions

To evaluate an expression means to find out the value of the expression by replacing the variable with a number and carrying out the calculations. If we evaluate an expression for many values of the variable we can produce a table of values.

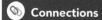 **Connections**

You evaluated some expressions for patterns in Chapter 6.

In Section 7.2.2, we simplified the expression for the weight, W, of pencils in Worked example 7.2 to $20x + 70$. Complete the table of values for $x = 20, 30, 40$ and 50.

x	20	30	40	50
W	$20 \times 20 + 70 = 470$			

The total weight of the box of pencils was 1570 grams. Look at the values in your table. Can you make an estimate about how many pencils were in each bag?

Algebraic expressions can contain more than one variable. Can you apply the same rules to evaluate expressions when there are many variables?

Explore 7.5

At Al's Café, a sandwich costs €2.50, a fizzy drink costs €1.50 and an ice cream costs €2.00. For a party, you need to purchase x sandwiches, y drinks and z ice creams.

Can you write down an expression for A, the total cost of your purchase at Al's Café?

Can you find out how much would you spend if $x = 7$, $y = 12$ and $z = 9$?

At Jodi's Deli, a sandwich costs €2.00, a fizzy drink costs €1.75 and an ice cream costs €2.50.

Can you decide where to shop for the party: Al's Café or Jodi's Deli?

We have to be careful when we evaluate expressions for negative values of variables.

Worked example 7.3

Evaluate $w = 3x - 2y$ when $x = -2$ and $y = -4$

Solution

To prevent mistakes, we can use brackets when we replace variables with numbers:

$w = 3x - 2y = 3 \times (-2) - 2 \times (-4)$

The product of two negative numbers is positive, so the two terms in the expression above become:

$3 \times (-2) = -6$

and

$-2 \times (-4) = 8$

So, $w = -6 + 8 = 2$

 Practice questions 7.2.3

1   The table shows the values of the variables a, b, c, x, y and z.

a	b	c	x	y	z
3	−1	0	2	1	4

Evaluate:

a $a + b$

b $b - c$

c $x^2 - b$

d $3x + a$

e $\dfrac{x}{3} - y$

f $-b - 3z$

g $\dfrac{y + x}{a}$

h $ax + b$

i $ax + by + c$

j $ax + by + cz$

k $\dfrac{cx + bz}{a}$

7.2.4 Brackets and expanding

 Explore 7.6

The two shapes below are made with pieces of string. The length of each side, in centimetres, is shown.

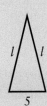

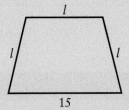

Can you write the length of string needed to make each shape using algebraic notation? Use brackets to group the length needed for each shape.

Can you write an expression for the total length of string needed to make both shapes? Do you still need brackets? How can you make the expression simpler?

What does 2 × 5 mean if we write it as repeated addition?

This can be either 2 + 2 + 2 + 2 + 2 or 5 + 5

What about 2 × (5 + 3)?

This is simply (5 + 3) + (5 + 3) or 2 fives + 2 threes, i.e. 2 × 5 + 2 × 3

Now look at these identical shapes, made with lengths of string.

We can write the length of string needed to make each shape using algebraic notation.

We may write the expressions for the two identical shapes as:

$(2x + 3)$ and $(2x + 3)$

Now, find an expression for the total length of string needed to make both shapes.

By collecting like terms, we can see that:

$(2x + 3) + (2x + 3) = 2x + 3 + 2x + 3 = 2$ lots of $(2x)$ and 2 lots of 3

$$= 2 \times 2x + 2 \times 3$$

We can also simplify the expression in a different way by realising that we have two lots of $(2x + 3)$:

$(2x + 3) + (2x + 3) = 2(2x + 3)$

Since the two quantities represent the total length of string:

$2(2x + 3) = 2 \times 2x + 2 \times 3$

$$= 4x + 6$$

 Fact

When we have a quantity multiplied by an expression in brackets, we need to multiply every term in the expression by the quantity outside the brackets. This is called **expanding the brackets**. So: $a(b + c) = ab + ac$.

Explore 7.7

Building on what you have learned, can you find the total length of string required to make the following shapes?

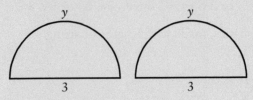

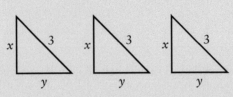

Can you find how much string you would need for four triangles? For 10 triangles? For m triangles?

 Fact

To expand brackets, use the formula:

$a(b + c) = ab + ac$

This is called the distributive property of multiplication over addition. This formula works the other way round too. If two terms have a common factor, we can put the common factor outside the brackets:

$ab + ac = a(b + c)$

This process is called **factorising**.

 Reminder

Remember, $2x - 4y$ is one way of writing $2x + (-4y)$

Worked example 7.4

a Expand the expression $3(x + 4)$

b Expand the expression $-3(2x - 4y + 5)$

Solution

a We need to remove the brackets from the expression $3(x + 4)$

We can apply the result $a(b + c) = ab + ac$ directly. We multiply the number 3 by both x and 4:

$$3(x + 4) = 3 \times x + 3 \times 4$$
$$= 3x + 12$$

b In this question, we need to be careful about the negative signs. We need to multiply -3 by $2x$ and by $4y$:

$(-3) \times 2x = -6x$, $(-3) \times (-4y) = 12y$ and $(-3) \times 5$

So, $-3(2x - 4y) = -6x + 12y - 15$

1   Write down the number of factors in each expression.

    a ab b $2wt$

    c $3x(x + 1)$ d $x(x - 2)$

    e $ab(c + 1)$ f $(a + c)(a + b)$

2   Expand these expressions.

    a $3(b + 3)$ b $2(w + t)$

    c $7(x - 1)$ d $(3 - x) \cdot 2$

    e $-2(x + 1)$ f $2(-c + 3)$

3   Expand these expressions.

    a $3(2b + c + 3)$ b $2(w + 2t - 3)$

    c $4(3s + 5t + 1)$ d $-2(2x - y + 1)$

    e $(2 + 3x)4$ f $2(3w - 1)$

4   Expand these expressions.

    a $b(b + 3)$ b $2t(w + t)$

    c $4e(e - 2)$ d $-x(x + 1)$

    e $3x(1 - 3x)$ f $-2y(3y + 1)$

5   Expand these expressions, then simplify them by collecting like terms.

    a $3(2b + 3) - 4$ b $2(1 + 2t) - t$

    c $3 + 4(2m - 1)$ d $x + 2(x - 1)$

    e $4(-1 - x) + 2x$ f $4(-y + 2) + 4y$

7.3   Indices

Algebraic expressions with multiplications often involve powers of the same variable. These can also be simplified.

Remember that an expression with a power, such as 3^4, means four 3s multiplied together: $3^4 = 3 \times 3 \times 3 \times 3$

Explore 7.8

Can you tell how many times x is multiplied by itself in the expression x^2? What about in the expression x^5?

Write down $x^2 \times x^5$ using multiplication signs. Can you write the result of $x^2 \times x^5$ as a single power of x?

Can you write these expressions as a single power of the base?

a $x^3 x^5$ **b** $a^2 a^4$ **c** $y \times y^2$

How would you write the product $x^m x^n$ as a single power of x?

Reminder

In a power expression x^a, x is the base.

When a fraction contains a power in the numerator and a power in the denominator, and both have the same base, we can simplify the fraction like this:

$$\frac{x^5}{x^3} = \frac{x \cdot x \cdot x \cdot x \cdot x}{x \cdot x \cdot x} = \frac{\cancel{x} \cdot \cancel{x} \cdot \cancel{x} \cdot x \cdot x}{\cancel{x} \cdot \cancel{x} \cdot \cancel{x}} = \frac{x^2}{1} = x^2$$

However, we also have $x^{5-3} = x^2$

When the base of a power is a power itself, we can simplify the expression like this:

$$(x^2)^3 = (x^2) \cdot (x^2) \cdot (x^2) = (x \cdot x) \cdot (x \cdot x) \cdot (x \cdot x) = x \cdot x \cdot x \cdot x \cdot x \cdot x = x^6$$

However, we also have $x^{2 \times 3} = x^6$

Fact

$x^m \cdot x^n = x^{m+n}$

$\dfrac{x^m}{x^n} = x^{m-n}$

$(x^m)^n = x^{mn}$

Practice questions 7.3

1 Write each expression to include a base raised to a single power.

  a $x \cdot x \cdot x$ **b** $w \times w \times w \times w \times w$

  c $x \cdot x \cdot x \cdot y$ **d** $x \cdot x \cdot y \cdot y$

  e $e \times t \times e \times t \times e \times t$ **f** $ab \times ab \times ab$

  g $x \cdot x \cdot x \cdot x \cdot y$ **h** $a \times a \times b \times a \times a \times b$

2 Expand these expressions, using multiplication signs instead of powers.

  a r^2 **b** x^3 **c** ab^2 **d** $(ab)^2$

  e xyz^3 **f** $(xyz)^3$ **g** $a^2 b^3$ **h** $x(yz)^2$

3 Write each expression to include a base raised to a single power.

a $r^2 r^3$ b $x^3 x$ c $ab^2 b^3$ d $a(ab)^2$

e $m^2 n^2 m^3 n$ f $3x^2 x(2y)$ g $2a(3b)^2 a^2$ h $n^2 m^3 nm^2$

4 Write each expression as a base raised to a single power.

🏆 Challenge Q4

a $\dfrac{r^3}{r^2}$ b $\dfrac{x^4}{x^2}$ c $\dfrac{y^2 \cdot y^3}{y^4}$ d $(x^2)^4$

e $(x^3)^3$ f $t^2 \cdot (t^2)^3$ g $\dfrac{(x^2)^3}{x^3}$ h $\dfrac{x \cdot (x^2)^3}{x^2 \cdot x^3}$

7.4 Algebraic equations

Explore 7.9

There are some pencils in the bag. If we know that there are 5 pencils in total, how many pencils are in the bag? What is the number x that makes $x + 1$ equal to 5? What is the same about these two questions?

Applied mathematicians use equations to model real-life situations, such as how far the ash will spread when a volcano erupts. By solving the equations, they can make accurate predictions that help people know what action to take.

Explore 7.10

Each bag contains the same number of pencils. If we know that there are 10 pencils in total, how many pencils are in each bag? What is the number x that makes $2x$ equal to 10? What is the same about these two questions?

An equation is an expression that contains an equal sign, such as $3 + 5 = 8$ or $20x + 70 = 1570$. Solving an equation means finding the value of the variable that makes the left hand side of the equation equal to the right hand side.

◎ Fact

The expression to the left of the equal sign is called the left hand side (LHS) and the expression to the right of the equal sign is called the right hand side (RHS).

Worked example 7.5

a Solve $x + 3 = 8$ **b** Solve $3x = 12$

Solution

a We need to find the number that we can add to 3 to get 8.

$x + 3 = 8$

The opposite of adding is subtracting. If we subtract 3 from both sides of the equation, we will be left with x on the left hand side.

$x + 3 - 3 = 8 - 3$

We can then evaluate the right hand side to find the value of x.

$x = 5$

If we replace x in the original equation with 5 we get

$5 + 3 = 8$, which is true.

b We need to find the number that we can multiply by 3 to get 12.

$3x = 12$

The opposite of multiplication is division. If we divide both sides of the equation by 3, we will be left with x on the left hand side.

$\dfrac{3x}{3} = \dfrac{12}{3}$

We can then evaluate the right hand side to find the value of x.

$x = 4$

If we replace x in the original equation with 4 we get

$3 \times 4 = 12$, which is true.

> **Fact**
>
> When we are working with equations, we need to do the same thing to each side of the equation to keep it balanced.

To solve equations that have both additions and multiplications, we need to take two steps.

 Worked example 7.6

Solve the equation $3x + 4 = 10$

Solution

$3x + 4 = 10$

We need to carry out operations on both sides of the equation until we are left with x by itself on one side.

Start by subtracting 4 from both sides.

$3x + 4 - 4 = 10 - 4$

$3x = 6$

Now divide both sides by 3.

$$\frac{3x}{3} = \frac{6}{3}$$

$$x = 2$$

The solution is $x = 2$

If we replace x in the original equation with 2 we get

$3 \times 2 + 4 = 10$, which is true.

Now we can go back to the situation in Explores 7.9 and 7.10 and find out how many pencils are in each bag.

Worked example 7.7

A box of pencils has been delivered to a classroom.

The pencils have been packaged in two different ways:
- There are seven loose pencils.
- There are two identical bags, each containing the same unknown number of pencils.

The label on the box says, 'weight of one pencil = 10 grams, total shipping weight = 1570 grams'.

How many pencils are in each bag?

Solution

We have already written and simplified an equation to represent the information.

$10x + 10x + 70 = 1570$, where x is the number of pencils in each bag.

As before, we need to carry out operations on both sides of the equation until we are left with x by itself on one side.

First subtract 70 from both sides. Then divide both sides by 20.

$$10x + 10x - 70 = 1570 - 70$$

$$20x = 1500$$

$$\frac{20x}{20} = \frac{1500}{20}$$

$$x = 75$$

So, there are 75 pencils in each bag.

To check that we have solved the equation correctly, we can substitute our value for x back into the equation and evaluate it:

$20 \times 75 + 70 = 1570$

This matches the total weight given for the box, so we can be confident that our answer is correct.

 Reflect

What would you do if your answer for the number of pencils was $x = 3.64$ or $x = 124\,000$? Are these reasonable answers in the context of the question?

We can use the same process to solve equations involving subtractions and divisions. Remember that the opposite of subtraction is addition and the opposite of division is multiplication.

What is the number x that makes $x - 1$ equal to 5?

 Worked example 7.8

a Solve the equation $x - 3 = 4$ **b** Solve the equation $\dfrac{x - 7}{9} = 2$

Solution

a The opposite of subtraction is addition, so add 3 to both sides.

$$x - 3 = 4$$
$$x - 3 + 3 = 4 + 3$$
$$x = 7$$

b The opposite of division is multiplication, so start by multiplying both sides by 9.

$$\frac{x - 7}{9} = 2$$
$$9 \times \left(\frac{x - 7}{9}\right) = 9 \times 2$$
$$x - 7 = 18$$
$$x - 7 + 7 = 18 + 7$$
$$x = 25$$

🛡 Hint

Write each step in your equation neatly. As you get more confident in the process, you might leave out some of the interim steps.

1 Write out in words the question that corresponds to each equation, then solve it.

   a $x + 1 = 3$ b $x - 3 = 6$

   c $3 + x = 4$ d $x - 3 = -2$

   e $6 - x = 2$ f $4 + x = 6$

   g $3x = 12$ h $4x = -16$

   i $21 = 7x$ j $-2x = 8$

2 Solve these equations.

   a $x + 3 = 6$ b $p + 3 = 4$

   c $x + 3 = 10$ d $m + 7 = 15$

   e $3 + w = 12$ f $2 + x = 2$

   g $15 + d = 18$ h $24 + m = 31$

   i $x - 6 = 2$ j $p - 4 = 2$

   k $y - 3 = 8$ l $x - 12 = 9$

   m $13 - x = 8$ n $4 - x = 2$

   o $10 - x = 4$ p $21 - x = 16$

   q $3x = 21$ r $2x = 28$

3 Solve these equations.

   a $x + 3 = -6$ b $-p + 3 = 4$

   c $x - 3 = -10$ d $m - 7 = 15$

   e $3 + w = -12$ f $-2 + x = 14$

   g $15 + d = -18$ h $-24 + m = -31$

   i $x - 6 = -2$ j $p - 4 = -2$

   k $-y - 3 = 8$ l $-x - 12 = -9$

   m $13 - x = -8$ n $-4 - x = 2$

   o $10 - x = -4$ p $-21 - x = 16$

   q $3x = -21$ r $-2x = 28$

4 Solve these two-step equations.

   a $2x + 1 = 9$ b $2x + 1 = 10$ c $3x - 1 = 11$

   d $2x + 5 = 23$ e $9 - 2x = 11$ f $5x - 10 = 10$

   g $3x + 4 = 4$ h $24 - 4x = 0$ i $6x - 10 = 14$

5   Solve these equations.

a $x + 1.3 = 4.4$ b $x - 3.2 = -3.4$ c $4.3 - x = 1$

d $1.2x = 3.6$ e $5x = 37.5$ f $1.3x = 13$

g $x + 1.4 = 0.8$ h $2x - 1 = 1.5$ i $1.5x + 3 = 1.5$

j $\dfrac{x}{2} + 1 = 2$ k $3 - \dfrac{x}{3} = 2$ l $\dfrac{x}{2} - \dfrac{3}{2} = 5$

6   Expand the brackets then solve these equations.

a $3(x + 1) = 6$ b $2(1 - x) = 4$

c $2(1 - x) = 5$ d $10(3 + x) - 9x = 31$

e $x + 3(x - 1) = 0$ f $3 + 8(2x - 1) - 14x = 2$

g $2 - (x + 3) = 4x + 3 - 4(x - 1)$ h $x - (x - 3) = 2x + 3$

Hint 6g

$2 - (x - 3)$ means
$2 + (-1) \times (x - 3)$

 7.5 **Problem-solving skills**

7.5.1 Word problems

 Explore 7.11

Some birds migrate every autumn from Iceland to West Africa, a distance of 6000 km. Geolocators used to track the migration of such birds show that they can complete the non-stop sea crossing flying at estimated speeds of 50 km per hour.

Can you set up an equation that helps you find an estimate of how long it takes such birds to fly this distance?

In this section, we are going to look at how algebra can be used as a problem-solving tool.

Worked example 7.9

A brick weighs a kilogram plus half the weight of the brick. What is the weight of the brick?

Solution

In this question, the unknown quantity is the weight of the brick. We can use the variable w to represent this quantity.

We have been given only one piece of information: the weight of the brick equals one kilogram plus half the weight of the brick.

We can convert this statement into an equation and then solve it.

the weight of the brick	w
equals	$=$
one kilogram	1
plus	$+$
half the weight of the brick	$\frac{1}{2}w$

So, our equation is $w = 1 + \frac{1}{2}w$

Now we need to solve this equation.

There is a w on both sides of this equation. We do not yet have a method for solving an equation with an unknown value on both sides, but we can use a table to evaluate both sides for different values of w.

LHS, w	0	1	2	...
RHS, $1 + \frac{1}{2}w$	1	1.5	2	...

The LHS equals the RHS when $w = 2$, so this is the solution to our problem: the brick weighs 2 kilograms.

Two kilograms is a reasonable weight for a brick, so our answer makes sense in the context of the question.

Reflect

Look at the two different wordings used in Worked example 7.9.

- A brick weighs a kilogram plus half the weight of the brick.
- The weight of the brick equals one kilogram plus half the weight of the brick.

What is the same about these sentences? What is different? Which one is easier to understand? Which one is easier to change into an equation?

We can apply these steps to other word problems.

1 Reword the given information so it contains the word 'equals'.

2 Introduce a variable for the unknown quantity.

3 Convert the given information into expressions for the left and the right hand sides of an equation.

4 Set up and solve the equation.

5 Check that our answer is reasonable.

Practice questions 7.5.1

1 Choose a suitable variable or variables and convert each of these into an algebraic expression.

   a   three times the mass

   b   4 metres more than the perimeter

   c   the sum of two consecutive numbers

   d   twice Annalisa's age

   e   two years older than Fred

   f   one year younger than Bill

   g   the difference between the base and the height

   h   one half of the product of base and height

   i   one half of the sum of the two diagonals.

> **Hint Q1c**
>
> Consecutive numbers come one after the other. For instance, 13 and 14 are consecutive numbers because 14 is the next number after 13.

2 Find the number that satisfies each statement by writing and solving an equation.

   a   A number plus 4 is 10.

   b   Subtracting 5 from a number gives 7.

   c   The difference between a number and 4 is 1.

   d   Three times a number is 12.

   e   A number divided by 4 is 6.

   f   Twenty divided by a number is 5.

   g   A number plus 5 is −3.

   h   Five minus a number is 0.

   i   The difference between 4 and a number is −1.

   j   Five times a number is 2.

   k   A number divided by 3 is −4.

   l   Negative ten divided by a number is 5.

3 Find the number that satisfies each statement by writing and solving a two-step equation.

   a   The difference between three times a number and 8 is 1.

   b   Four plus twice a number is 8.

   c   Dividing a number by three and adding 8 gives 10.

   d   Subtracting four times a number from 50 gives 30.

4 Solve each problem by writing and solving an equation.

   a   A set of notebooks costs $32. Each notebook costs $4. How many notebooks are in the set?

   b   Another set of notebooks costs $25. There are 5 notebooks. How much does each notebook cost?

c Three kilograms of apples and two kilograms of pears cost €12. One kilogram of pears costs €3. How much does one kilogram of apples cost?

d The sum of two consecutive whole numbers is 11. Find the two numbers.

e The sum of two consecutive even numbers is 22. Find the two numbers.

f The sum of three consecutive odd numbers is 39. Find the three numbers.

g Three sandwiches and two drinks cost $12.00. A sandwich costs twice as much as a drink. Find the cost of a drink.

h Three sandwiches and two drinks cost $12.00. A sandwich costs $0.50 more than a drink. Find the cost of a drink.

5 A supercomputer requires 24 modules of RAM to operate. Each RAM module is 20 gigabytes. How many 16 gigabyte modules could be used instead?

7.5.2 Graphical solutions

 Explore 7.12

Construct a table of values for $y = x + 5$ for $x = 0, 1, 2, 3, 4$. Plot your values on a graph.

Construct a table of values for $y = 7 - x$ for $x = 0, 1, 2, 3, 4$. Plot your values on the same graph.

What do you notice about the points you have plotted? What does this tell you about the equation $x + 5 = x + 7$?

 Connections

You learned how to draw graphs of rules in Chapter 6.

Graphs can be a powerful method for solving equations.

 Worked example 7.10

The sum of three consecutive whole numbers is twice the largest number. Find the smallest number of the three.

Solution

We can follow the steps from the end of Section 7.5.1.

1 Reword the given information so it contains the word 'equals'.
The sum of three consecutive whole numbers equals twice the largest number.

2   Introduce a variable for the unknown quantity.

The value we need to find is the smallest number of the three. We will call this number x.

We do not need to choose variables for the other two numbers, because we can express these numbers in terms of x. The next consecutive whole number is 1 more than x, so it is $x + 1$, and the third is $x + 2$.

3   Convert the given information into expressions for the left and the right hand side of an equation.

On the left hand side, we have 'the sum of three consecutive whole numbers'. This is:

$x + (x + 1) + (x + 2)$

We can simplify this to $3x + 3$

On the right hand side, we have 'twice the largest number'. This is:

$2(x + 2)$

We can expand the brackets to get $2x + 4$

4   Set up and solve the equation.

To set up the equation, we put an equals sign between the left hand side and the right hand side.

$3x + 3 = 2x + 4$

We can use a table of values to solve an equation like this. Another method is to use graphs.

We start by constructing a table of values for each side of the equation, for the same values of x.

x	0	1	2	3	...
LHS, $3x + 3$	3	6	9	12	...

x	0	1	2	3	...
RHS, $2x + 4$	4	6	8	10	...

Now we can plot these points on the same graph. We use dots for the LHS points and crosses for the RHS points, so that we can tell them apart.

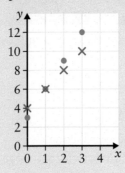

Looking at the graph, there is one point that has both a dot and a cross: (1,6). This is the point where $x = 1$ and $y = 6$

This means that when $x = 1$, both the left hand side and the right hand side of the equation are equal to 6. So, the solution to $3x + 3 = 2x + 4$ is $x = 1$

The smallest number of the three consecutive whole numbers is 1.

5   Check that our answer is reasonable.

If 1 is the smallest of the three numbers, then the other two are 2 and 3, since they are consecutive.

The sum of the three numbers is $1 + 2 + 3 = 6$

Twice the largest number is $2 \times 3 = 6$

These results are the same, so this confirms that our answer is correct.

 Reflect

Which method do you prefer: constructing a table of values or drawing a graph? Can the graphical method be used for any equation? What steps do you need to follow to use the graphical method?

 Investigation 7.3

There are many graphing software packages available for free online. With the help of your teacher, librarian or guardian, find some of these packages and learn how to use them.

 Research skills

Use a software package to solve the equations $\dfrac{1}{x} = x$ and $x^2 = x$

 Practice questions 7.5.2

1   Create a table of values for both the LHS and the RHS of each equation. Then graph both the LHS and the RHS on the same Cartesian plane and estimate a solution to the equation.

a $3x + 1 = 9 - x$ b $3 - 2x = x - 6$

c $x + 5 = 10 - x$ d $3(x + 1) = -2(x - 9)$

e $2x - 1 + 3(x - 1) = 2x + 2$ f $x + 5 = 10$

g $0 = -3 + x$ h $2(x + 3) = 3x + 6$

i $3 + 2(x - 4) = x - 5$ j $3x - 1 = -x - 9$

k $x + 1 = -5 - x$ l $3(x - 4) + 1 = 2(x + 1) + x$

m $3(x - 4) + 14 = 2(x + 1) + x$

2   a   Plot graphs for each of the following equations for $x = 0, 1, 2, 3, 4$.
     i $2(x - 1) = 2x + 2$ ii $3x + 1 = 3(x - 4) + 13$

     b   Describe how many solutions there are to each of the equations in part a.

 Hint Q1j

The solution can be negative. Use the trend shown by the graph to choose which x values you need for the table of values.

 Hint Q2

When no x value makes the LHS equal to the RHS, the equation has no solution. In this case, we say that the equation is inconsistent or unsolvable. When any x value makes the LHS equal to the RHS, the equation has infinitely many solutions. In this case, we say that the equation is indeterminate.

⭐ Self assessment

○ I can use the conventions and abbreviations used for algebraic expressions.	○ I can expand expressions with brackets, for both positive and negative numbers.
○ I can distinguish between terms and factors.	○ I can simplify products of powers.
○ I can use index notation.	○ I know what solving an equation means.
○ I can recognise and use grouping symbols.	○ I can solve linear equations in one and two steps.
○ I can evaluate and simplify algebraic expressions.	○ I can convert a word problem into an equation.
○ I can recognise the meaning of algebraic expressions in context.	○ I can solve linear equations graphically.

? Check your knowledge questions

1   Expand the brackets, then simplify each algebraic expression.

a $3(x - 1) + 4(x + 5)$ b $2(1 - x) - 4(x + 4)$

c $4(2x - 1) + 3(3 - x)$ d $-3(x - 2) + 3(3 - 2x)$

e $-2(x - 1) + 3(-x + 3)$ f $2(x - 2) - 2(x - 2)$

g $a - b + 3a + 4b$ h $a(3 + b) - b(2 + a)$

i $xy - 3x(y + 1) - x$ j $x^2 + 3x - x(x + 1)$

k $x(x + 1) - 2x(2x - 1)$ l $ax(x + y) - x(ax - ay)$

2   Evaluate each expression when $x = 1$, $y = 2$, $a = 0$ and $b = -1$

a $3x + y$ b $x^2 - 3y$ c $a + b + 2x$

d $xy + 3$ e $\dfrac{y}{x} + b$ f $ax + y$

g $\dfrac{x + y}{3} - a$ h $a^2 + bxy$ i $b(x + y) + b$

j $a(x - b) + y^2$ k $\dfrac{bx - y}{y} - \dfrac{ay - x}{x}$ l $x\dfrac{b}{y} + b\dfrac{x}{y}$

3   Solve each equation.

a $x - 2 = 3$ b $r + 4 = 12$ c $3 + w = 5$ d $10 = x - 7$

e $4 = 7 - x$ f $3 + x = 5$ g $3 + x = 3$ h $3 + x = 1$

i $3 + x = 0$ j $-x - 4 = 3$

4   Solve each equation.

a $3x = 12$ b $4x = 8$ c $15 = 5x$

d $\dfrac{x}{2} = 8$ e $\dfrac{x}{3} = 4$ f $\dfrac{x}{6} = 5$

g $2x = -6$ h $-2x = 10$ i $-x = 10$

j $\dfrac{x}{2} = -4$ k $\dfrac{x}{-3} = 7$ l $\dfrac{x}{-2} = -4$

m $\dfrac{21}{x} = 7$ n $\dfrac{16}{x} = -4$ o $\dfrac{14}{-x} = 7$

5   Simplify and solve each equation.

   a $3x + 1 = 13$ b $-3 + 2x = 9$

   c $2x - 1 = -9$ d $4x - 1 = -5$

   e $3(x + 1) - x = 11$ f $x + 2(x + 3) = 3$

   g $10 = 3x + 4(x - 1)$ h $3x - 2(x - 1) = 4$

6   An equilateral triangle has sides of length x. The perimeter of the triangle is 24 cm. Find the lengths of its sides.

7   The base of a rectangle is 12 cm longer than its height. The perimeter of the rectangle is 60 cm. Find the base of the rectangle.

8   The base of a rectangle is twice the height. The perimeter of the rectangle is 60 cm. Find the base and the height of the rectangle.

9   The sum of two consecutive whole numbers is 55. Find the two numbers.

10   Frank's phone costs 1.2 times as much as Andrew's phone. The total value of Frank and Andrew's phones is $220. How much does Andrew's phone cost?

11   The amount of money M (in dollars) earned by Jacques when he works for x hours is given by the rule $M = 3x + 5$

   a   How much does Jacques earn when he has worked for 4 hours?

   b   Jacques wants to buy a new T-shirt that costs $35. How many hours does he have to work to earn enough money?

12   A school bus company charges its customers £50 a year for the subscription to its service. Each trip to and from the school costs an additional £0.50.

   a   Carolina uses the bus a total of 234 times. How much does the company charge her?

   b   The company charges Laura £153. How many times did she use the bus?

13 A class of 30 students is planning on buying a present for their teacher's birthday, but only 25 students want to participate in the present. If all the students in the class participated, the contribution would be $5.00 each.

Find how much each of the 25 students has to pay.

14 The nuclei of different elements on the periodic table contain a different number of protons. Hydrogen has one, helium has two, lithium has three and beryllium has four.

p represents a single proton, so a beryllium nucleus can be drawn like this:

In nuclear reactions, nuclei are split and the protons in them are recombined in different arrangements: a proton and a helium nucleus could combine and give a lithium nucleus. This reaction is represented as:

For each reaction below, find the name of the nucleus with the unknown number, z, of protons.

a

b

c

15 Solve the following equations by graphing both the left and the right hand sides on the same Cartesian plane.

a $x + 1 = 3 - x$ b $4 - x = x + 2$ c $2x - 4 = 5 - x$

d $5 + x = 2x$ e $5 + x = -2x - 1$ f $2(x + 4) = 3x + 8$

Measure

8

8 Measure

 KEY CONCEPT

Relationships

 RELATED CONCEPTS

Generalisation, Quantity, Space, Systems

 GLOBAL CONTEXT

Orientation in space and time

Statement of inquiry

Using systems of measure to investigate the relationships between different quantities in space and time allows us to solve complex real-life problems.

Inquiry questions

Factual

- How do you find the perimeter of
  - a triangle
  - a quadrilateral
  - a regular polygon?
- What is the formula for the area of
  - a rectangle
  - a parallelogram
  - a triangle?
- What is the formula for the volume of a cuboid?

Conceptual

- When might it be useful to know
  - the perimeter of an object
  - the area of an object
  - the volume of an object?
- Why is it useful to generalise relationships using algebra?

Debateable

- Do all countries and cultures use the same standard measures for length and time? What are the advantages or disadvantages?

Do you recall?

1   Name these quadrilaterals

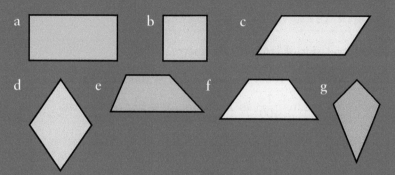

2   What is the mathematical name for each 3D shape?

3   This blue rod has length x.

a   Write an expression for the length of four of these rods.

This yellow rod has length y.

b   Write an expression for the length of three of these rods.

c   Write an expression for the total length of two blue rods and five yellow rods.

4   How many days are there in:

a   a week   b   a normal year   c   a leap year?

5   How many hours are there in a full day (from midnight to midnight).

6   How many seconds are there in one minute?

7   How many minutes are there in one hour?

8   Write these 24-hour times using a.m. or p.m.

a   0800   b   1700   c   1930   d   1228   e   0015

9   Write these times in the 24-hour clock.

a   11.15 a.m.   b   1.40 a.m.   c   3.10 p.m.   d   10.30 p.m.

10 Write 'quarter to 7 in the evening' as:

a   a 12-hour time using a.m. or p.m.

b   a 24-hour time.

8.1 Measuring lengths and perimeters

8.1.1 Metric units of length

The basic metric unit of length is the metre.

The prefix 'kilo' means 1000, so 1 kilometre = 1000 metres.

The prefix 'centi' means $\frac{1}{100}$, so 1 centimetre = $\frac{1}{100}$ of a metre.

The prefix 'milli' means $\frac{1}{1000}$, so 1 millimetre = $\frac{1}{1000}$ of a metre.

You can use these abbreviations for the units of length.

kilometre	km
metre	m
centimetre	cm
millimetre	mm

Explore 8.1

How can you use the definitions above and measuring instruments to help you complete these conversions?

1 m = ___ mm

1 m = ____ cm

1 cm = ____ mm

1 km = ___ m = ____ mm

Worked example 8.1

This ruler is marked in millimetres.

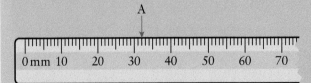

Write the measurement shown at A:

a in millimetres

b in centimetres and millimetres

c in centimetres.

Solution

a 32 mm   We can read the value directly from the mm scale.

b 3 cm 2 mm   10 mm = 1 cm, so 30 mm = 3 cm

c 32 ÷ 10 = 3.2 cm   To convert mm to cm, divide by 10.

 Reflect

How could you write the length 4 cm 6 mm in two other ways?
How do you write 2 mm in cm?

Practice questions 8.1.1

1   This ruler is marked in millimetres.

Write down the measurements shown by the arrows:

    a   in millimetres   **b**   in centimetres.

2   Here is part of a ruler marked in centimetres.

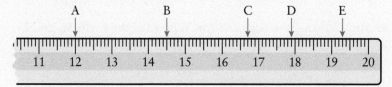

Write down the measurements shown by the arrows:

    a   in centimetres   **b**   in millimetres.

3   Complete these conversions from larger units to smaller units.

    a   7 cm = ___ mm   **b**   4.8 cm = ___ mm

    c   142 cm = ___ mm   **d**   3 m = ___ cm

    e   14 m = ___ cm   **f**   2.5 m = ___ cm

    g   2 m = ___ mm   **h**   1.6 m = ___ mm

    i   26 km = ___ m   **j**   5.7 km = ___ m

4 Complete these conversions from smaller units to larger units.

    a 49 mm = ___ cm   b 120 mm = ___ cm

    c 400 cm = ___ m   d 96 cm = ___ m

    e 3000 mm = ___ m   f 5000 m = ___ km

    g 375 m = ___ km   h 300 m = ___ km

5 Complete these measurement conversions.

    a 36 m = ___ cm   b 490 cm = ___ m

    c 0.7 m = ___ cm   d 230 cm = ___ mm

    e 12 km = ___ m   f 1.2 m = ___ mm

    g 2700 m = ___ km   h 4750 cm = ___ m

6 Write these lengths in order, from longest to shortest.

    5470 mm   5.2 m   0.01 km   574 cm   0.005 km

7 Rishi is 170 cm tall. What is his height in metres?

8 A swimming pool is 25 m long. How many lengths of the pool make 1 km?

9 How many shelves 40 cm long can be cut from a piece of wood 2.5 m long?

10 Petra runs $2\frac{3}{4}$ km. Sam runs 2650 m. Who runs the furthest, and by how much?

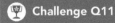 **Challenge Q11**

11 A description of a fridge states:

> Dimensions (cm) H94 × W54.5 × D59.5

Will this fridge fit:

    a into a space 520 mm wide

    b under a worktop of height 900 mm?

 Challenge Q12

12 The height of the Eiffel Tower in Paris is 300 m.

Pierre wants to build a scale model of the Eiffel Tower, one twentieth of the height of the real tower. Work out the height of Pierre's model, in centimetres.

Investigation 8.1

Does every country use the metric system? What units are commonly used to measure length in your country?

Research the units used to measure length in other countries.

Engineers want to build a railway from the United States across the border into Canada. Why might the different measurement systems used by these two countries cause some problems when deciding the width of the track?

How could these problems be resolved?

 Research skills

8.1.2 Units, accuracy and rounding

 Explore 8.2

Think about the different objects you could measure, and the smallest unit you could measure, with different measuring instruments.

You could record your ideas in a table like this:

Measuring instrument	Smallest unit you can measure with it	Examples of objects you can measure with it
15 cm ruler	mm	Pencil, badge, mobile phone

 Fact

Microscopically small objects are measured in micrometres (one millionth $\frac{1}{1\,000\,000}$ of a metre) and nanometres (one thousand-millionth $\frac{1}{1\,000\,000\,000}$ of a metre).

The accuracy of a measurement depends on the instrument used to measure it.

Rulers and tape measures are marked in centimetres and/or millimetres to measure short lengths. One revolution of a trundle wheel measures 1 metre, for measuring longer distances.

Trundle wheels are used to measure longer distances.

The odometer in a car measures the distance travelled to the nearest tenth of a kilometre. The tenths digit is shown in red. This odometer shows that the total distance travelled by the car since it was made is 81 057.6 km.

You need to choose suitable units to measure an object. You would not measure the thickness of a book in metres, or the distance across a field in centimetres.

Sometimes you do not need a very accurate measurement. When you measure a person's height, a measurement to the nearest centimetre is probably accurate enough.

Laser rangefinders are used by surveyors to make accurate measurements.

Worked example 8.2

A laser rangefinder measures the dimensions of a room as 3.746 m by 3.873 m.

Give the dimensions in metres, to the nearest 10 centimetres.

Solution

3.746 m = 374.6 cm

374.6 rounds to 370 to the nearest 10

370 cm = 3.7 m

3.873 m = 387.3 cm

387.3 rounds to 390 to the nearest 10

390 cm = 3.9 m

The dimensions are 3.7 m by 3.9 m, to the nearest 10 cm.

Reflect

In Worked example 8.2, how did it help to convert the measurements to cm?

Can you round 5.26 m to the nearest 10 cm without converting to cm first?

Worked example 8.3

Marta wants to cut 1 metre of ribbon into three equal lengths.

How long should each length be?

Give your answer to a suitable degree of accuracy.

Solution

$100 \div 3 = 33.333...$ cm

It is not possible to measure 33.333… cm accurately.

Each length is 33.3 cm (to the nearest millimetre).

Reminder

Always state how you have rounded the measurement in the final answer.

Reflect

What would you use to measure the length of each piece of ribbon?

What units would use? Why is 33.3 cm of a suitable degree of accuracy?

1   Write 7.517 metres as ___ m ___ cm ____ mm

2   Write 3 m 24 cm 5 mm:

   a   in centimetres, to the nearest mm

   b   in metres, to the nearest mm.

3   Here is a section of a 5-metre tape measure, marked in millimetres.

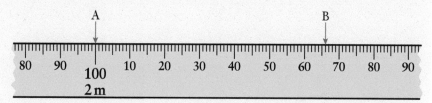

Write down the measurements marked with arrows, in metres, to the
nearest millimetre.

4   For each measurement, write down:

   i   suitable units

   ii   a suitable instrument to take the measurement.

   a   the length of a car

   b   the height of a teddy bear

   c   the width of a phone screen

   d   the width of a door

   e   the height of a room

   f   the length and width of a sports pitch

   g   the distance from your home to the airport.

5 Measure these lines to the nearest centimetre.

a

b

c

d

e

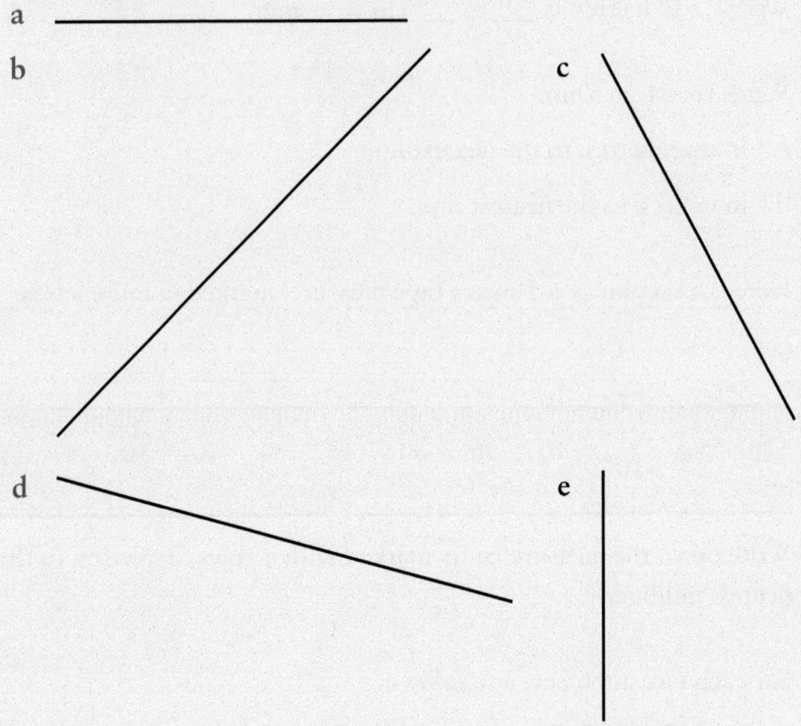

6 Measure the lines in question 5 to the nearest millimetre.

7 A laser rangefinder measures the dimensions of a room as 2.675 m by 3.120 m.

Give the dimensions in metres, to the nearest 10 centimetres.

8 A football pitch has length 106.24 m and width 68.75 m. Write these measurements accurate to the nearest tenth of a metre.

9 The width of a doorway is one third of its height.

The height is 2 metres.

Work out the width, to a suitable degree of accuracy.

10 A running track round a park is 386 m long. Shona runs 7 laps of the track every day for a week. How far does she run in total? Give your answer in km, to the nearest 100 m.

11 Which of these measurements round to 9 cm, to the nearest centimetre?

8.4 cm   8.5 cm   8.7 cm   8.9 cm   9.1 cm   9.4 cm   9.5 cm

Challenge Q12

12 A wall is 2.3 m long, to the nearest 10 cm. A sofa is 228 cm long. Explain why the sofa may be longer than the wall.

13 Here is a car odometer, with the tenths digit shown in red.

5 3 0 9 6 | 4

a Write the total distance travelled in km.

b Raj says, 'After another 3600 m the car will have travelled exactly 53 100 km.'

Explain why Raj may not be correct.

 Investigation 8.2

Measure the heights of several doors in your home or school. Work out a suitable estimate for the height of a door.

 **Research skills**

Use your estimate for the height of a door to help you estimate the heights of the people in this picture.

Find everyday objects that you can use to help you estimate other lengths. For example the width of your thumbnail may be approximately 1 cm.

8.1.3 Perimeter of polygons and rectilinear shapes

The perimeter of a shape is the distance around the outside of the shape. You can measure the perimeter of a shape, or you can calculate it, if you know the lengths of its sides.

Sometimes, you have to work out the lengths of some of the sides first, using the properties of the shape.

Explore 8.3

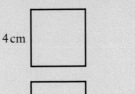

This square has side length 4 cm. Can you find the perimeter of this square?

What if the square has side length l? Can you express the perimeter in terms of l?

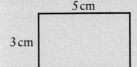

This rectangle has length 5 cm and width 3 cm. Can you find the perimeter of this rectangle?

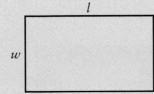

What if a rectangle has length l and width w? Can you express the perimeter in terms of l and w?

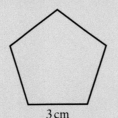

This regular pentagon has side length 3 cm. Can you find the perimeter of this regular pentagon?

How would you express the perimeter of a regular pentagon with side length l in terms of l?

How would you express the perimeter of a regular hexagon with side length l in terms of l?

How would you express the perimeter of a regular n-sided polygon with side length l in terms of l?

A rectilinear shape has straight edges that meet at right angles.

You can use opposite sides of a rectilinear shape to work out any unknown lengths on the shape.

Work out the perimeter of this rectilinear shape.

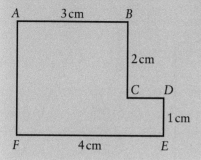

Solution

Understand the problem

All the vertices are right angles. Not all the side lengths are given.

Make a plan

We need to find the unknown side lengths, using the lengths given. Then add together all the side lengths to find the perimeter.

Carry out the plan

First we find the unknown side lengths.

$AF = BC + DE$
$\quad = 2 + 1$
$\quad = 3\,\text{cm}$

$AB + CD = FE$
$\quad 3 + CD = 4$
$\qquad CD = 4 - 3$
$\qquad\quad = 1\,\text{cm}$

Now we can add the lengths going around the shape, starting from A.

Perimeter $= 3 + 2 + 1 + 1 + 4 + 3$
$\qquad\qquad = 14\,\text{cm}$

Look back

How many sides does the shape have? We can check that we added the lengths of all the sides.

The shape has 6 sides. The sum for the perimeter includes 6 values.

 Reflect

Does it matter which of the unknown side lengths you find first?

Practice questions 8.1.3

1 Measure the side lengths of these shapes to the nearest millimetre, then calculate their perimeters.

a 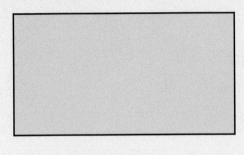 b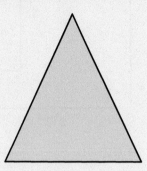

c 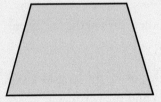 d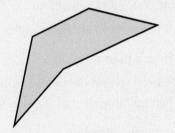

2 Find the perimeter of this shape to the nearest:

   a millimetre

   b centimetre.

3.2 cm

2.1 cm

3 These shapes are drawn on a centimetre square grid. Find the perimeter of each shape.

   a

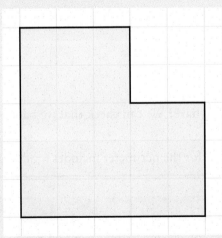

b

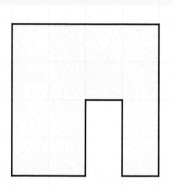

4   Find the perimeter of:

    a   an equilateral triangle with sides measuring 5.3 cm

    b   a regular hexagon with sides measuring 4 cm

    c   a regular nonagon with sides measuring 6 cm

    d   a regular octagon with sides measuring 15 mm.

5   The perimeter of a square is 49 cm. Find the length of one side of the square.

    Give your answer to a suitable degree of accuracy.

6   A regular nonagon and a regular pentagon both have perimeter 54 cm. Find the difference in their side lengths.

7   Work out the perimeters of these rectilinear shapes. All measurements are in metres.

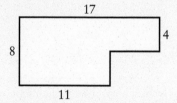
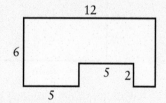

8   Find the perimeter of this shape.

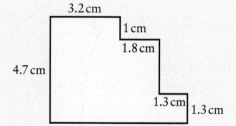

9 Calculate the perimeter of this quadrilateral.

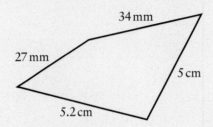

10 In this triangle, $AB = BC$ and $AC = \frac{1}{2}BC$

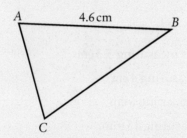

Work out the perimeter of the triangle.

Challenge Q11

11 Two sides of a kite are 6 cm and 9.5 cm.
Find the perimeter of the kite.

Challenge Q12

12 The diagram shows the measurements of a field.

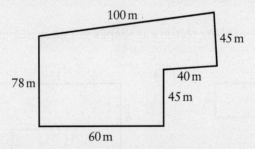

Hint Q12

You would need to
purchase a whole number
of 10-metre rolls, as
they cannot be split into
smaller lengths.

Wire fencing comes in 10-metre rolls.
One 10 metre roll of fencing costs $15.99
Calculate the cost of fencing for the field.

Challenge Q13

13 The diagram shows a regular pentagon and
a square joined together.
The total perimeter of the combined shape
is 98 cm.
What is the perimeter of the square?

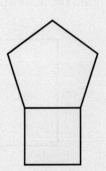

8.2 Area

8.2.1 Area of rectangle and parallelogram

The area of a 2D shape is the amount of space inside the shape. Area is measured in square units, some of which are shown here.

centimetres squared, or square centimetres cm²	millimetres squared, or square millimetres mm²	metres squared, or square metres m²
1 cm² 1 cm 1 cm	1 cm 1 mm²	1 m² 1 m 100 cm = 1 m

Explore 8.4

This shape is drawn on a centimetre squared grid.
Can you find its area?

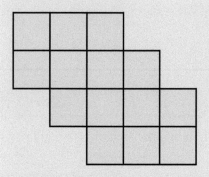

Communication skills

Could you estimate the area of this irregular shape?

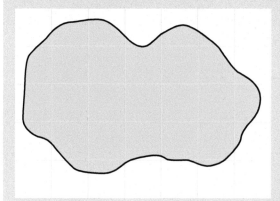

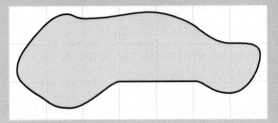

Worked example 8.5

Estimate the area of this irregular shape.

Solution

Count all the whole squares, and all the squares that are *more than* half inside the shape.

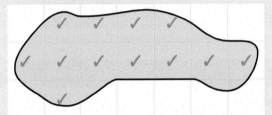

The question does not tell you the size of the squares, so give the answer in square units or units^2.

Area $\approx$ 12 square units

Reflect

How useful is 'ticking' the squares when finding the area of an irregular shape?

Investigation 8.3

 Thinking skills

This rectangle is drawn on a centimetre square grid.

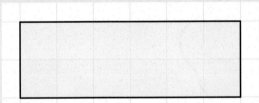

There are 2 rows of 6 squares inside the rectangle.
The area of the rectangle is $6 \times 2 = 12$ cm^2.

1 Write multiplication calculations for the areas of these rectangles:

a    b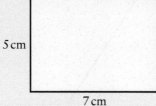

2 Generalise your result from question 1 to write an expression for the area of a rectangle with length l and width w.

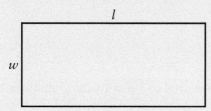

Hint b

'Generalise' your result means 'write a general statement that applies to all rectangles'. You can use algebra to write the statement.

3 Explain why you can calculate the area of a square in the same way as you calculate the area of a rectangle.
Write an expression for the area of a square with side l cm.
What is the connection between the square numbers and the area of a square?

4 Draw a parallelogram like this on squared paper.

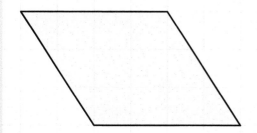

Cut out your parallelogram. Cut off a right-angled triangle from one end, as shown on the next page.

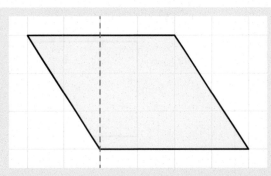

Move the triangle to the other end of the parallelogram.
What shape does this make?

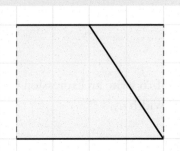

Does this happen for every parallelogram?

5   Which two measurements do you need to know to calculate the area
    of a parallelogram?

6   Write an expression for the area of a parallelogram.

The area of a rectangle
is:

> Area = length ×
> width
> $A = l \times w$

Or, $A = lw$

◴   **Worked example 8.6**

Work out the area of this shape.

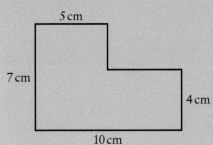

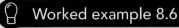

Solution

There is no formula for calculating the area of an L shape.

We need to split the shape into rectangles, so we can use the area of a rectangle formula. Then we find the unknown lengths we need to work out the areas of the rectangles.

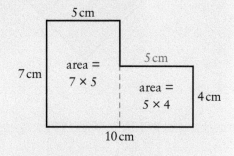

Total area = 7 × 5 + 5 × 4

= 35 + 20

= 55 cm

Fact

The area of a parallelogram is:

Area = base × height

$A = b \times h$

Or, $A = bh$

Reflect

Could you split the shape in Worked example 8.6 into rectangles in another way? Would you get the same answer?

Hint

The height is vertical and the base is horizontal. So, the height is perpendicular to the base.

Practice questions 8.2.1

1. These shapes are drawn on a centimetre squared grid. Find the area of each shape.

a 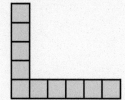 b 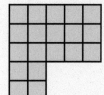 c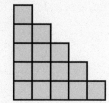

2. The outer square of each diagram has side length 1 cm. What fraction of a square centimetre is the shaded area?

a   b   c   d

3. Find the area of each shape drawn on a 1 cm squared grid.

a
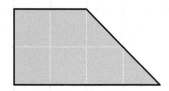

b

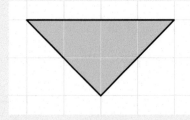

c

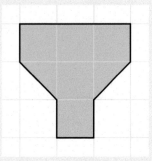

4   This circle is drawn on 1 cm squared grid paper. Estimate the area of the circle.

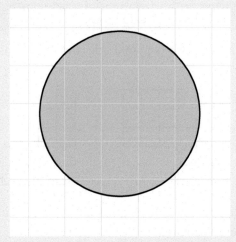

5   Find the area and perimeter of this shape drawn on a centimetre squared grid.

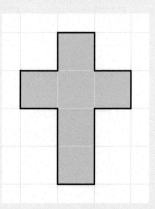

6 Calculate the area of each rectangle.

a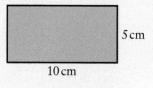
5 cm
10 cm

b
10 mm
12 mm

c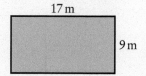
17 m
9 m

7 Calculate the area of a square with side 7 cm.

8 A square has area 121 cm². What is the side length of the square?

9 The area of this rectangle is 27.2 cm².
   Work out the width of the rectangle.

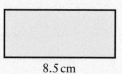
8.5 cm

10 Calculate the area of each rectangle in the units given.

a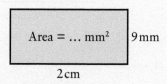
Area = … mm²   9 mm
2 cm

b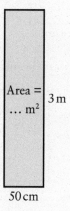
Area = … m²   3 m
50 cm

c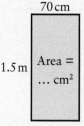
70 cm
1.5 m   Area = … cm²

Hint Q10a

To calculate an area in mm², make sure the length and width are both in mm.

11 Calculate the area of each parallelogram.

a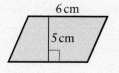
6 cm
5 cm

b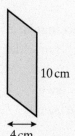
10 cm
4 cm

c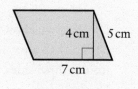
4 cm   5 cm
7 cm

12 Find the distance between the two parallel sides of this parallelogram.

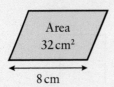

13 Calculate the area of each shape.

a

3 cm
8 cm
5 cm
4 cm

b

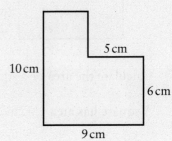

5 cm
10 cm
6 cm
9 cm

c

12 mm
2 mm
7 mm
4 mm

d

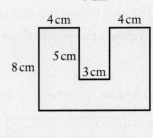

4 cm   4 cm
5 cm
8 cm   3 cm

Challenge Q14

14 Work out the shaded area in each shape.

a

8 cm
4 cm
10 cm

b

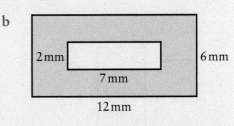

2 mm   6 mm
7 mm
12 mm

Challenge Q15

15 A rectangular room measures 3.2 m by 2.8 m.
Carpet costs $13 per square metre.
Work out the cost of carpet for the room.

Challenge Q16

16 The diagram shows a rectangular
garden with a paved path 1 m wide
around three sides.
The rest of the garden is lawn.
Work out the area of the lawn.

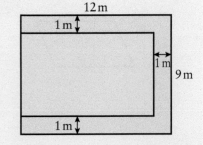

12 m
1 m
1 m
9 m
1 m

17 Find the dimensions of the rectangle with perimeter 36 cm that has the largest possible area.

Challenge Q17

18 Find the dimensions of the rectangle with area 18 cm² that has the largest possible perimeter.

Challenge Q18

8.2.2 Area of a triangle

Explore 8.5

a Draw a triangle inside a rectangle like this.

Cut out the red triangle.

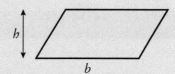

Try to fit the two smaller triangles on to the shaded triangle. What do you notice?

What is the area of the rectangle?

What do you think the area of the red triangle is?

b Draw a parallelogram like this.

Join opposite vertices with a line to make two triangles.

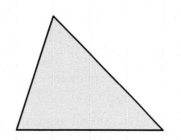

What is the area of the parallelogram?

What do you think the area of one of the triangles is?

c Make two copies of this triangle on squared paper and cut them out.

Can you rearrange them to make a parallelogram? Is there more than one parallelogram you can make?

Repeat with this triangle.

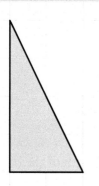

Worked example 8.7

Calculate the area of this triangle.

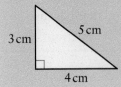

Solution

We start by writing down the formula for the area of a triangle.

$A = \frac{1}{2}bh$

Then we substitute the values for the base b and the height h.

$A = \frac{1}{2} \times 4 \times 3$

$= 2 \times 3$

$= 6 \text{ cm}^2$

Worked example 8.8

Calculate the area of this shape.

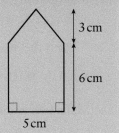

Solution

There is no formula for calculating the area of this shape.

We need to split the shape into a rectangle and a triangle, so we can use those area formulae.

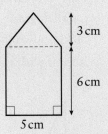

Total area = $6 \times 5 + \frac{1}{2} \times 5 \times 3$

$= 30 + 7.5$

$= 37.5 \text{ cm}^2$

Reflect

What property of rectangles do you use to find the length of the base of the triangle?

Practice questions 8.2.2

1   Calculate the area of each triangle.

a

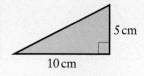

5 cm

10 cm

b

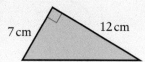

7 cm   12 cm

c

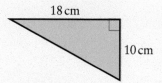

18 cm

10 cm

d

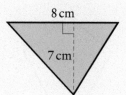

8 cm

7 cm

e

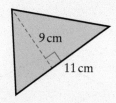

9 cm

11 cm

f

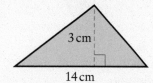

3 cm

14 cm

2   Work out the area of each triangle.

 Hint Q2

The height of a triangle is the perpendicular distance from a vertex to the opposite side.

In these triangles, this height is shown outside the triangle.

a

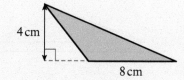

4 cm

8 cm

b

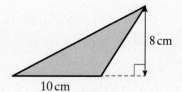

8 cm

10 cm

c

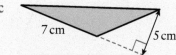

7 cm   5 cm

3   Find the area and perimeter of this triangle.

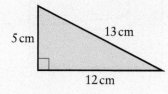

5 cm   13 cm

12 cm

4   Calculate the area of these shapes.

a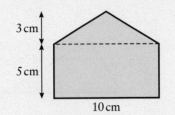

b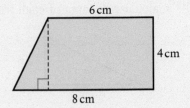

c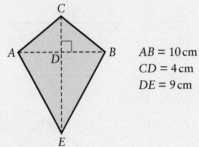

$AB = 10\,cm$
$CD = 4\,cm$
$DE = 9\,cm$

5   Calculate the area of this rhombus.

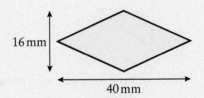

6   The area of this triangle is $14\,cm^2$. Find its height.

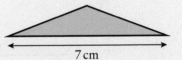

7   Calculate the area of this trapezium.

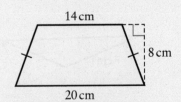

8   Work out the area of the shaded shape.

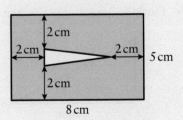

9   Calculate the area of this shape.

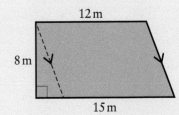

10 The diagram shows the shape of a wooden floor.

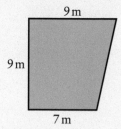

9 m

9 m

7 m

The floor needs to be painted with varnish.

Varnish costs $16 for a 5-litre can.

One litre of varnish covers 3 m².

Work out the cost for one coat of varnish.

11 Each vertex of the shaded square is the midpoint of a side of the outer square.

🏆 Challenge Q11

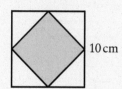

10 cm

Calculate the area of the shaded square.

12 Make accurate drawings of three different right-angled triangles, each with an area of 16 cm².

🏆 Challenge Q12

🔗 Connections

You learned how to draw triangles accurately using a ruler and protractor in Chapter 4.

8.3 Volume

The volume of a 3D shape is the amount of 3D space inside the shape.
Volume is measured in cube units.

centimetres cubed, or cubic centimetres cm³	millimetres cubed, or cubic millimetres mm³	metres cubed, or cubic metres m³
 1 cm / 1 cm / 1 cm	 1 cm / 1 mm³	1 m / 100 cm = 1 m / 1 m

Explore 8.6

How can you find the volume of this shape made from 1 cm cubes?

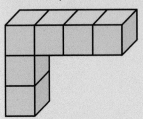

Each of the following cuboids is made from centimetre cubes.

Can you find the volume of each cuboid?

a

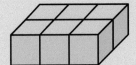

b

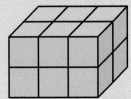

c

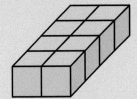

d

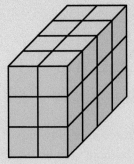

How can you work out the number of cubes in the top layer of this cuboid?

When you know the number of cubes in one layer, how can you find the number of cubes in the cuboid?

How can you work out the number of cubes in this cuboid?

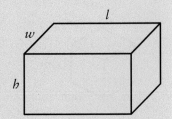

Calculate the volume of this cuboid.

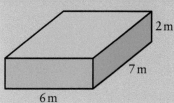

 Fact

The formula for the volume of a cuboid is:

Volume = length × width × height

$$V = lwh$$

Solution

$V = lwh$	Write the formula.
$V = 6 \times 7 \times 2$	Substitute the measurements.
$V = 84 \, m^3$	The measurements are in metres, so the units of volume are m^3.

 Reflect

Does it matter if you multiply the length, width and height in a different order?

 Practice questions 8.3

1 Each of these solid shapes is made from centimetre cubes.

Find the volume of each solid shape by counting cubes.

a 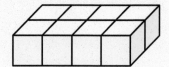   b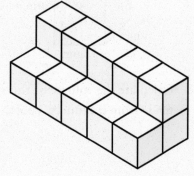

c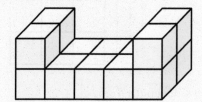

2 Work out the volume of each cuboid in cubic centimetres.

a

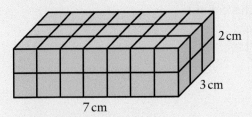

2 cm
3 cm
7 cm

b

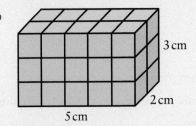

3 cm
2 cm
5 cm

3 Calculate the volume of each cuboid.

a

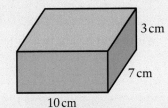

3 cm
7 cm
10 cm

b

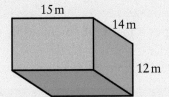

15 m
14 m
12 m

c

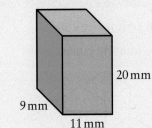

20 mm
9 mm
11 mm

4 Calculate the volume of this cube made from centimetre cubes.

5 Find the volume of a cube with side length 5 cm.

6 a Sketch and label a cuboid with length 4 cm, width 3 cm and height 5 cm.

   b Calculate the volume of the cuboid.

7 Calculate the volume of this cuboid in cm³.

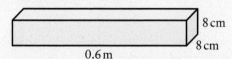

🛡 **Hint Q7**

Convert all the lengths to cm first.

8 Calculate the volume of this cuboid in m³.

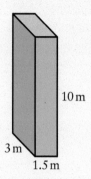

9 This cuboid has volume 60 cm³.
Work out its height.

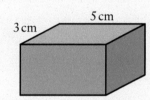

10 A cube has volume 343 mm³. What is the length of one side of the cube?

🏆 **Challenge Q10**

11 A child's toy box is a cuboid. Its dimensions are height 40 cm, length 72 cm and width 38 cm.

Show that the volume of the toy box is greater than 100 000 cm³.

🏆 **Challenge Q11**

12 Find the volume of each shape made from cuboids.

🏆 **Challenge Q12**

a
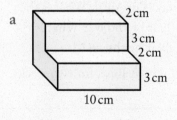

b
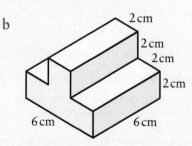

13 A cereal box is a cuboid with base 20 cm by 7 cm and height 25 cm.
The cereal in the box fills the box to a height of 22 cm.
Work out the volume of empty space in the box.

🏆 **Challenge Q13**

🛡 **Hint Q13**

Make a sketch.

Challenge Q14

14 a   Sketch and label the measurements of three different cuboids with volume $24\,cm^3$.

    b   Sketch and label another cuboid with volume $24\,cm^3$ where at least one of the side lengths is not a whole number of cm.

 Challenge Q15

15   Find the volume of your classroom.

Challenge Q16

16   What is the side length of the largest cube you can make with 200 centimetre cubes?

 8.4   **Time**

8.4.1 Time calculations

 Explore 8.7

Use a calendar for this year to help you answer these questions.

On what day of the week is your birthday this year?

Ask three students their birthdays. Which of these birthdays is closest to yours? How many days is it between the two birthdays?

Which birthday is furthest from yours? How many days is it between the two birthdays?

Kyle's birthday is in February. Lara's birthday is in April. What is the greatest possible number of days between their birthdays? What is the smallest possible number of days between their birthdays?

Jayden is going on holiday on the first Saturday in June, for two weeks. What date does his holiday end?

 Fact

Some countries write dates in day/month/year format:

   10/3/20

   10th day, 3rd day of 10th month, in 2020

   10 March 2020

Other countries write dates in month/day/year format:

   10/3/20

   10th month, 3rd day in 2020

   3 October 2020

How do you write the date in your country? Which format do you prefer and why?

We use measures of time to state when an event has happened or is supposed to happen:

   School starts at 8.40 a.m.

We also use them to measure the duration of an event:

   The school day is 6 hours and 25 minutes long.

Worked example 8.10

Sanjay is planning a holiday starting on 19 June and ending on 15 August.

How many days is the holiday?

Solution

We need to work out the number of days in each whole month or part of a month, then find the total.

We count 19 June as day 1, 20 June as day 2, and so on.

19 June to 30 June = 12 days

July = 31 days

1 August to 15 August = 15 days

Total = 12 + 31 + 15 = 58 days

 Reflect

Why do you need to know the number of days in each month to answer the question in Worked example 8.10?

Would a timeline like this help you to work out the answer?

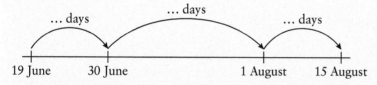

 Worked example 8.11

Macey leaves home at 10.40 a.m. and walks for $\frac{3}{4}$ hour to her friend's house.

What time does she arrive at her friend's house?

Solution

$\frac{3}{4}$ hour is 45 minutes.

We can draw a timeline to split the 45 minutes into 20 minutes up to 11.00 a.m., and then 25 minutes after.

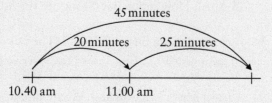

Macey arrives at 11.25 a.m.

✏ Practice questions 8.4.1

1 What date is:

   a three days after 9 March

   b one week before 9 March

   c two weeks after 9 March?

2 What date is two weeks before December 5th?

3 A teacher gives her students an assignment on 23rd October.
   The students have two weeks to work on the assignment.
   What date is the assignment due to be handed in?

4 Copy and complete.

   a $\frac{1}{2}$ hour = ☐ min

   b $\frac{1}{4}$ hour = ☐ min

   c $\frac{3}{4}$ hour = ☐ min

   d $\frac{1}{2}$ minute = ☐ seconds

   e 1 hour 20 minutes = ☐ minutes

   f 225 minutes = ☐ hours ☐ minutes

   g 98 seconds = ☐ min ☐ seconds

5 How many:

   a minutes in a day   b seconds in an hour   c hours in a week?

6 Sam works a 40-hour week. What fraction of the total hours in a week
   is this?

7 The school holidays start on 22 March. The next term starts on
   15 April. How many days' long are the holidays?

8 Darcy catches a bus at 2.55 p.m. The bus journey into town takes
   20 minutes. What time does Darcy arrive in town?

9 A train leaves London at 9.36 a.m. and arrives in York $2\frac{1}{2}$ hours later.
   What time does the train arrive in York?

10 The table shows the start and end times of three films at a cinema.

Film	Start time	End time
HORROR MOVIE	12.40 p.m.	2.30 p.m.
ADVENTURE PICTURE	3.30 p.m.	5.35 p.m.
FUNNY FILM	6.20 p.m.	8.45 p.m.

a   Which film is the longest?

b   How many minutes shorter is the shortest film than the longest film?

11 Serena makes a cake. She puts it in the oven at 11.15 a.m. The cake needs to bake for 50 minutes. When will the cake be ready?

12 Pranav is roasting a chicken for lunch. The chicken will take 1 hour 40 minutes to roast. Lunch is at 1.15 p.m. What time should Pranav put the chicken in the oven to roast?

13 A ship leaves an island at 4.10 p.m. on Tuesday. It arrives at the mainland at 3.30 a.m. on Wednesday. How long did the ship's journey take from the island to the mainland?

14 Jodie catches a bus to school. The bus journey takes 35 minutes.
Buses leave the stop near Jodie's home at 10 minutes or 40 minutes past every hour.
Jodie needs to arrive at school before 8.40 a.m.
What time should she catch the bus?

15 A teacher is planning an activity day. The day will start at 9 a.m. and finish at 3 p.m.
There need to be morning and afternoon breaks of at least 10 minutes each.
There needs to be a lunch break of at least 40 minutes.
Here are the activities, and the times needed for each.

Challenge Q15

Art	1 hour
Drama	2 hours
Swimming	45 minutes
Pottery	45 minutes in the morning and 1 hour in the afternoon
Dance	40 minutes
Singing	25 minutes
Poetry	$\frac{3}{4}$ hour
Creative writing	90 minutes

Plan a timetable for the day, fitting in a range of activities.
Show the start and end time of each activity.

8.4.2 Timetables

Timetables often use 24-hour times, so they can show morning and afternoon times clearly.

Here is part of a train timetable for a route in the South West of England.

Plymouth	1518	1654
Saltash	1527	1703
Par	1610	1846
Truro	1634	1910

There are two trains from Plymouth to Truro, one at 1518 and one at 1654. Each train stops at Saltash and Par on the journey.

Here is part of a bus timetable for buses from the Station to the Market Square.

Station	0715	0735	0755	0815
Main Street	0718	0738	0758	0818
Park	0723	0743	0803	0823
School	0727	0747	0807	0827
Market Square	0733	0753	0813	0833

The 7.35 a.m. bus from the station arrives at the park at 7.43 a.m.

Worked example 8.12

Use the bus timetable above to work out:

a   how long the journey takes from Main Street to Market Square

b   which bus to catch from the station to arrive at the school between 8.00 a.m. and 8.15 a.m.

Solution

a

Station	0715
Main Street	0718
Park	0723
School	0727
Market Square	0733

We can use any of the four bus journeys shown in the timetable to work out the journey time from Main Street to Market Square.

The bus leaves Main Street at 0718 and arrives at Market Square at 0733. We can draw a timeline to help find how long the journey takes.

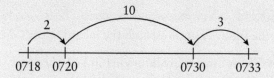

2 + 10 + 3 = 15 minutes

It takes 15 minutes to get from Main Street to Market Square.

b The times in the question are given as 12-hour times. In 24-hour time, they are 0800 and 0815.

We need to look along the row for the school to see which bus arrives between 0800 and 0815.

Station	0715	0735	0755	0815
Main Street	0718	0738	0758	0818
Park	0723	0743	0803	0823
School	0727	0747	0807	0827
Market Square	0733	0753	0813	0833

Then look at the top of the column to find the time this bus leaves the station.

Catch the bus that leaves the station at 0755. In 12-hour time, this is 7.55 a.m.

Worked example 8.13

How long is the journey from Saltash to Truro?

Plymouth	1518	1654
Saltash	1527	1703
Par	1610	1846
Truro	1634	1910

Solution

The train leaves Saltarsh at 1527 and arrives at Truro at 1634. We can draw a timeline to help find how long the journey takes.

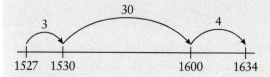

3 + 30 + 4 = 37 minutes

The journey from Saltarsh to Truro takes 37 minutes.

The meridian line through Greenwich has been the reference point for GMT since 1884.

Investigation 8.4

Find a time-zone map on the internet and research the meaning of 'GMT'.

If where you live is not on GMT, how many hours behind or ahead of GMT is it?

Which countries have more than one time zone?

What time is it in New York when it is 4 p.m. in London?

What time is it in San Francisco when it is 10 a.m. in New York?

What time is it in Oslo when it is 6 a.m. in Tokyo?

Worked example 8.14

A plane leaves London for Athens at 1540 GMT.

The flight takes 3 hours 45 minutes.

Athens time is 2 hours ahead of London time. What time does the plane arrive in Athens, local time?

Solution

We need to work out the time in London when the plane lands in Athens. Then work out what this time will be in local Athens time.

We can draw a timeline to split the time into chunks.

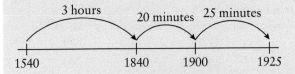

The plane arrives in Athens at 1925 London time.

Athens time is 2 hours ahead of London time, so we need to add on 2 hours.

The plane arrives at 2125 Athens time.

Reflect

You could solve Worked example 8.14 by working out the time in Athens when the plane leaves London, and adding the flight time to that. Which method do you prefer, and why?

1   Here is part of the TV schedule for one evening.

7 p.m.	Soap
7.30 p.m.	Quiz
8 p.m.	Documentary
9 p.m.	Drama
9.45 p.m.	News
10.15 p.m.	Weather
10.20 p.m.	Sport

a   Ali turns on the TV at 7.35 p.m. How long does he have to wait for the documentary?

b   How long are the news and weather programmes in total?

c   Which is the longest programme?

2   Here is a bus timetable.

Town centre	1536	1614	1700	1740
Hightown	1550	1628	1714	1754
Park	1602	1640	1726	1806
Rail station	1617	1655	1741	1821
Airport	1630	1708	1754	1834

a   How long is the bus journey from the town centre to the airport?

b   What is the longest time interval between stops? Which are the two stops?

c   How long is the journey from the park to the airport?

d   How long is the journey from Hightown to the rail station?

e   Susie lives in Hightown. She needs to be at the airport by 6.30 p.m. Which buses could she catch from Hightown?

3   The table shows the tide times for three days at a seaside resort.

Tuesday	Wednesday	Thursday
High 0153	High 0302	High 0400
Low 0843	Low 0936	Low 1022
High 1528	High 1611	High 1649
Low 2128	Low 2218	Low 2300

a   How long is it between the two high tides on Tuesday?

b How much later is the evening low tide on Thursday than on Tuesday?

c The sea is best for surfing two hours either side of high tide. Kai wants to surf on Thursday. Between what times should he go to the sea?

d Good fishing time is up to three hours either side of low tide. On Wednesday, Rhianna arrives at the sea at 1015. How much good fishing time does she have?

4 Here are the train timetables from Dortmund to Wuppertal, and Wuppertal to Dortmund.

Dortmund	0835	0935	1035
Hagen	0903	1003	1103
Wuppertal	0919	1019	1119

Wuppertal	1615	1700	1805
Hagen	1631	1716	1821
Dortmund	1659	1744	1849

Bella lives in Dortmund and wants to visit her grandmother in Wuppertal.

She catches the first train after 9 in the morning from Dortmund.

What is the longest she can stay in Wuppertal?

5 Here is a bus timetable:

HOWTH - DUBLIN - GREYSTONES Monday to Friday								
Howth	0700	0740	0850	0953	1030	1102	1115	1146
Sutton	0703	0744	0853	0956	1033	1105	1118	1149
Bayside	0705	0746	0856	0958	1035	1107	1120	1151
Howth Junction	0709	0750	0859	1002	1039	1111	1124	1155
Kilbarrack	0710	0751	0901	1003	1040	1112	1125	1156
Raherry	0712	0753	0903	1005	1042	1114	1127	1158
Harmonstown	0714	0755	0905	1007	1044	1116	1129	1200
Killester	0717	0758	0907	1010	1047	1119	1132	1203
Clontarf Road	0720	0801	0911	1013	1050	1122	1135	1206
Dublin Connolly arr	-	-	0914	-	-	1125	-	1210
dep	0724	0805	0925	1017	1054	1135	1139	-
Tara Street	0727	0808	0928	1020	1057	1136	1142	1212
Dublin Pearse arr	0728	0809	0921	-	1058	-	-	1214
dep	0734	0811	0931	1022	1106	1139	1144	-
Grand Canal Dock	0737	0813	0933	1024	1108	-	1146	1216
Lansdowne Road	-	0815	0935	1026	1110	-	1148	1218
Sandymount	-	0817	0937	1028	1112	-	1150	1220

a What time is the first bus from Tara Street to Sandymount?

b For how long does the 1030 bus from Howth wait at Dublin Pearse?

c How long does the last bus from Howth take to get to Grand Canal Dock?

d Flynn lives in Raheny. He needs to be at Tara Street by 10.30 a.m. It takes him 7 minutes to walk from his house to the bus stop in Raheny. What is the latest time he should leave the house?

e Sara gets to the bus stop in Howth at 11 a.m. She wants to go to Landsdowne Road. How long does she have to wait for her bus?

6 A plane leaves London for Istanbul at 1145 GMT.
The flight takes 4 hours 5 minutes.
Istanbul time is 2 hours ahead of London time. What time does the plane arrive in Istanbul, local time?

7 A flight leaves Sydney for Taipei at 0800.
The flight takes $9\frac{1}{2}$ hours.
The time in Taepei is 2 hours behind Sydney.
What time does the flight arrive in Taipei, local time?

8 A plane leaves New York at 1750 (local time) and arrives in Los Angeles at 2048 (local time). The time in Los Angeles is 3 hours earlier than the time in New York.
How long is the flight?

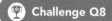 Challenge Q8

⭐ Self assessment

I can measure lengths accurately.

I can choose suitable units to measure length.

I can convert between metric measures of length.

I can round measurements of length to a suitable level of accuracy.

I can compare lengths given in different metric units.

I can calculate the perimeter of a shape.

I can calculate the perimeter of a regular polygon given its side length.

I can work out unknown lengths on a rectilinear shape and calculate its perimeter.

I can calculate an unknown length on a shape, given its perimeter.

I can calculate the area of a rectangle.

I can calculate the area of a parallelogram.

I can calculate the area of a triangle.

I can work out unknown lengths on a rectilinear shape and calculate its area.

I can calculate an unknown length on a shape, given its area.

I can estimate the area of an irregular shape drawn on a centimetre squared grid.

I can calculate the area of a shape made from rectangles, triangles and parallelograms.

I can calculate the volume of a cube.

I can calculate the volume of a cuboid.

I can calculate the volume of a shape made from cubes and cuboids.

I can calculate an unknown length on a cuboid, given its volume.

I can work out the number of days between two dates.

I can work out the date 2 weeks later than a given date.

I can work out the duration of an event from its start and end times.

I can work out the start time of an event, given its end time and duration.

I can work out the end time of an event, given its start time and duration.

I can use timetables with 12-hour or 24-hour clock times.

? Check your knowledge questions

1 Complete these measurement conversions.

   a 420 m = ☐ cm

   b 530 cm = ☐ m

   c 0.4 km = ☐ mm

   d 546 mm = ☐ cm

   e 2450 m = ☐ km

   f 3.4 m = ☐ mm

2 How many books of width 18 mm will fit on a shelf 1 m long?

3 Write down suitable units and a suitable instrument to measure:

   a the length of an ant

   b the height of a flagpole

   c the distance between two villages

   d the distance between two classrooms

   e how long it takes to do five star jumps.

4 Rae cuts a 2 m length of wood into seven equal pieces.
   How long is each piece?
   Give your answer to a suitable
   degree of accuracy.

5 The diagram shows a girl standing next to a tree.
   Estimate the height of the tree.

6   Work out the perimeter of each shape.

a

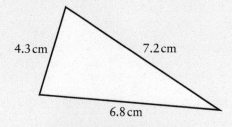

4.3 cm   7.2 cm

6.8 cm

b

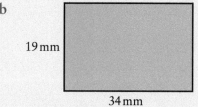

19 mm

34 mm

c

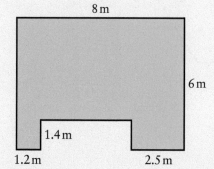

8 m

6 m

1.4 m

1.2 m   2.5 m

d

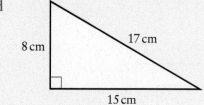

17 cm

8 cm

15 cm

7   Find the perimeter of a regular nonagon with sides measuring 7 cm.

8   Find the side length of a square with a perimeter of 10 cm.

9   This shape is drawn on a centimetre squared grid.

Estimate the area of the shape.

10 Calculate the area of this rectangle in square centimetres.

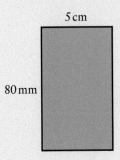

5 cm

80 mm

11 Calculate the area of each shaded shape.

a

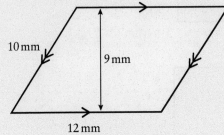

10 mm

9 mm

12 mm

b

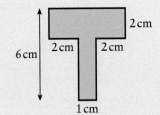

2 cm

2 cm   2 cm

6 cm

1 cm

c

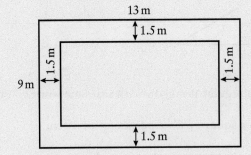

13 m

1.5 m

1.5 m   1.5 m

9 m

1.5 m

d

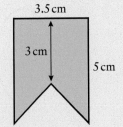

3.5 cm

3 cm

5 cm

12 The area of each shape is given. Find the height of each shape.

a

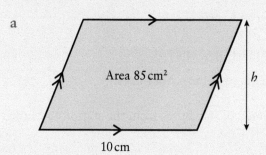

Area 85 cm²

h

10 cm

b

Area 56 cm²

h

14 cm

13 Find the volume of this cuboid.

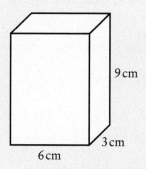

9 cm

3 cm

6 cm

14 This cuboid has a cuboid cut out through its centre.

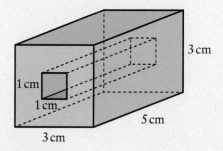

3 cm

1 cm

1 cm

3 cm

5 cm

Work out the volume of the remaining solid shape.

15 A train leaves Paris at 1527 and arrives in Lyon 1 hour 58 minutes later.

What time does it arrive in Lyon?

16 The school day starts at 0840 and ends at 1630.

How many minutes are there in the school day?

17 Here is part of a bus timetable for buses from the station to Market Square.

X1 from **Station** to **Market Square**				
Monday to Friday				
	X1	**X1**	**X1**	**X1**
Station	1215	1250	1325	1400
Main Street	1219	1254	1329	1404
Park	1226	1301	1336	1411
School	1231	1306	1341	1416
Market Square	1233	1308	1343	1418

a How often do the buses leave the station?

b Lee catches the 1254 bus from Main Street. What time does he arrive at the school?

c How long is the journey from the park to Market Square?

18 A flight leaves Hong Kong for Amsterdam at 0530.

The flight takes $12\frac{1}{2}$ hours.

The time in Hong Kong is 6 hours ahead of Amsterdam.

What time does the flight arrive in Amsterdam, local time?

Statistics

9

9 Statistics

 KEY CONCEPT

Form

 RELATED CONCEPTS

Quantity, Representation

 GLOBAL CONTEXT

Fairness and development

Statement of inquiry

Representing data in different forms allows us to identify patterns that can be used to help us make fair decisions.

Inquiry questions

Factual

- How do we gather useful information?
- How do we represent information?

Conceptual

- What type of data do we collect?
- How does the way we present data influence its understanding?

Debateable

- How do we use information from data to make effective decisions?
- What is the most appropriate way of presenting different types of data?
- What makes one data representation more helpful than another?

Do you recall?

1   Students chose their favourite colour.
    The bar chart shows the results.

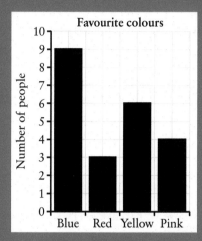

    a   How many students chose yellow?

    b   Which colour was the most popular?

2   Round each number to one decimal place.

    a   4.78   b   10.55

3   Work out the number halfway between:

    a   10 and 14   b   4 and 5.

4   Work out the size of angle x.

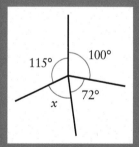

5   Work out:

    a $\frac{1}{2}$ of 46   b $\frac{5}{8}$ of 24.

6   In a group of 20 students, 13 of them are pescatarians.

    a   What fraction of these students are pescatarians?

    b   What percentage of these students are pescatarians?

 Fact

A pescatarian is someone
who eats fish, but does not
eat any other meat.

9.1 Collecting data

Mr Garry suspected that maths was the favourite subject of students in MYP1, so he asked students in his class to conduct a survey to verify this.

Catherine decided to ask five of her friends. Three of them liked maths the best, so she reported that 3 out of 5, or 60% of students, liked maths best.

Eric surveyed his class. 20 out of the 30 students liked maths best, so he reported that $\frac{2}{3}$ out of the grade were maths lovers.

Tim decided to ask 10 students from each of the five MYP classes for their opinion. From this, he concluded that 70% of students liked maths best.

Alex thought the only way to be really sure was to ask every student in the MYP group.

Which survey would be likely to give the most accurate results?

Why do you think the other surveys might have been less accurate?

Which survey would be likely to give the least accurate results?

Statistics is the branch of mathematics concerned with the gathering and organisation of information called **data.** There are three basic steps to this process:

- collecting the data
- arranging and presenting the data
- analysing the data and drawing conclusions.

In this chapter, we will discuss the first two of these.

Collecting data

Information is collected in different ways, depending upon how it is going to be used and how accurate it needs to be.

Collecting information from the whole population is a **census**.

In a census, data is collected about every individual in a population. For example, you might want to know how many students at your school take a bus to and from home. If you ask this question of every student at your school, this would be a **simple census.**

If a population is large, a census is impractical, so only a small portion or **sample** of the population is surveyed. These data are used to estimate the characteristics of the whole population.

Fact

In many countries a **national census** is conducted every ten years, collecting information about citizens such as details of income, employment, education, housing and many other things. This is a huge task, and it takes a lot of time and effort to sort and analyse the data.

When a sample is taken, it needs to be representative of the whole population. For example, the proportion of people below the age of 18 or the proportion of women and men in the selected sample should be the same as the proportions in the total population.

Samples should be selected **at random**. This means that every member of the population should have an equal chance of being chosen.

A sample must also be large enough to enable you to make conclusions. If you selected five students at random, and four of them caught a bus to school, the sample would be too small to reasonably conclude that 4 out of 5 of all students caught a bus to school. If the sample was 100 students, then any data collected would have a greater chance of reflecting the total student population.

The 2020 US census was the first US census to allow citizens to respond by phone or online, as well as by post.

◌ Worked example 9.1

Kevin attends an international school in Japan. He wants to know the most popular sport in the international community. He handed out this questionnaire to his fellow students.

Tick your favourite	
Soccer	[ ]
Rugby	[ ]
Hockey	[ ]
Basketball	[ ]
Cricket	[ ]
Tennis	[ ]
Softball	[ ]

After collecting the results, Kevin determined that soccer was by far the most popular sport.

a Describe one problem with Kevin's sample.

b How could Kevin make his sample more representative?

c Describe one problem with Kevin's questionnaire.

d Design a questionnaire that would address the problem you described in part c.

Solution

a We need to think about Kevin's sample and whether or not it represents the population he is interested in. The population is the international community in Japan. The sample is drawn from one international school.

Kevin's sample is biased and may not represent the whole community. It is limited to students at his own school, who would all have a similar range of ages and experiences.

b To make a sample representative of a population, we need to look at the characteristics of the population and make sure that they are the same in the sample.

Kevin could divide the community into different age groups and take a random sample from each age group.

c   We need to look at the questionnaire and work out if the data collected will answer the question 'What is the most popular sport?'.

There are several problems with the questionnaire.

- Not all sports are listed. Some people might have a favourite sport that is not on the list.

- The order of the list of sports might influence the result.

- Some people's favourite sport might depend on if they are playing the sport or watching the sport. Kevin's questionnaire does not capture this information.

d   To address the problems described, a questionnaire could ask people to list the three sports they like playing the most.

Reflect

What is one problem with conducting a census, rather than examining a sample?

Practice questions 9.1

1   Elli wanted to see which type of music was most popular in the community. She handed out this survey to students at her school. She discovered that rock and roll was the most popular.

Tick Your Favourite Music	
Classical	
Rock & Roll	
Jazz	
Folk	
Techno	

a   Describe one problem with Elli's sample.

b   How could Elli improve her sample?

c   Describe one problem with Elli's questionnaire?

d   How could Elli improve her questionnaire?

2   Ms Barber is the principal of an elementary school. She suspected that maths was the least favourite subject of students in her school. She asked some students from the local high school to conduct a survey to investigate.

David decided to ask seven students from the playground. Five of them liked maths, so he reported that 2 out of 7 did not like maths.

Sophie decided to survey her younger brother's class. 20 out of the class of 30 liked maths the least, so she reported that $\frac{2}{3}$ of students have maths as their least favourite subject.

Matteo decided to ask 10 students from each of the five grades their opinion. From this, he concluded that 60% of students did not like maths.

Michelle thought the only way to be really sure was to ask each student at the school.

a   Which student carried out a census?

b   Three students surveyed a sample of the school. Which sample is likely to have been the most accurate?

c   Why do you think the other two samples might have been less accurate?

3   Hakan rolled a dice and got these results:

2   5   6   5   6   1   3   6

Because more 6s were rolled than any other number, Hakan concluded that the dice must be biased towards 6s.

a   Is this conclusion necessarily correct?

b   What should Hakan do to make his conclusion more likely to be correct?

c   How does this question relate to taking a sample?

4   A local TV magazine ran an opinion poll to see which programmes and celebrities are the most popular. Why might the results not be accurate?

5   A survey is to be held at your school. It has been decided to use random sampling to sample students from Grades MYP1 to MYP5. The number of students in each grade is shown in the table. If 80 students are to be chosen, how many from each grade should be chosen?

MYP 1	MYP 2	MYP 3	MYP 4	MYP 5	Total
160	203	245	200	202	1010

 Hint Q5

In a random sample, every member of the population has an equal chance of being selected. For a representative random sample, the proportion of MYP1 students in the sample should be the same as the proportion of MYP1 students in the population, and similarly for the other grades.

 Challenge Q5

9.2 Categorising and sorting data

Think about the different ways you might group the people in your class. You might group them by:

- height
- hair colour
- number of pets owned
- favourite food.

What other examples can you think of?

Which of your groups have numerical categories? Which ones have descriptive categories?

List some possible values for the groups that have numerical categories. Is there anything different about the types of values you list?

Data are the pieces of information that you collect through your study. It is important to recognise the different types of data.

- **Qualitative data** is descriptive. There are two types of qualitative data:
  - **categorical data** groups data items with no implied order
  - **ordinal data** has a natural order.

- **Quantitative data** is numerical. There are two types of quantitative data:
  - **discrete data** can take only specific values, such as integers
  - **continuous data** can take any value on a continuous scale.

Generally, if you count something then it is discrete. If you measure it, then it is continuous.

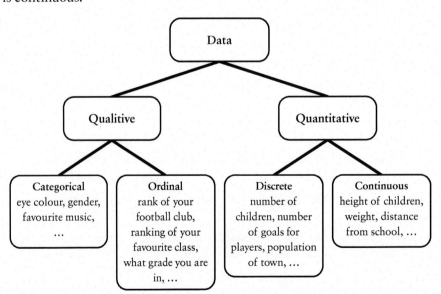

Worked example 9.2

Classify each of the following types of data.

a heights of members of the basketball team at your school

b nationalities of students at your school

c time students spend per week doing homework

d daily number of flu cases registered in a city during the last four weeks

e colours of cars parked next to school

f weights of teachers at your school

g ages, in years, of teachers at your school.

Solution

a Heights are measured, so this is continuous data.

b Nationality is descriptive, and there is no natural order to nationalities, so this is categorical data.

c Time is measured, so this is continuous data.

d Number of cases is counted, and must be an integer, so this is discrete data.

e Colour is descriptive, and there is no natural order to colours, so this is categorical data.

f Weight is measured on a continuous scale, so this is continuous data.

g Age in years is counted, so this is discrete data.

When a large amount of data has been collected, it needs to be sorted and presented in a form that is easily understood. A common method of sorting data is to use a **frequency distribution table**.

 Worked example 9.3

Amina asked the 30 students in her class how many pencils they bring to school each day. She collected the following data.

2	3	1	2	5	4	1	2	3	3
2	4	6	4	4	2	3	1	1	3
2	1	5	3	3	2	1	2	4	2

Create a frequency distribution table for the data and summarise the result.

Solution

We need to find out how many students bring one pencil, how many bring two and so on. Then we need to describe the main conclusions from the results.

We will create a table with a row for each number of pencils. We will use a tally to count how many students bring each number of pencils.

Outcome	Tally	Frequency
1	卌I	6
2	卌IIII	9
3	卌II	7
4	卌	5
5	II	2
6	I	1
	Total	30

The Outcome column shows the possible number of pencils.

The Tally column is used for recording the data, one entry at a time.

The Frequency is the total tally for each outcome (the number of students who bring that many pencils).

The most common number of pencils brought to school is 2.

Only one student brings 6 pencils to school.

There are six students who bring only 1 pencil.

The total of the frequencies is the same as the number of students Amina asked, so we know we have not missed any results.

Explore 9.3

A dice was rolled 36 times and the following results were recorded:

4	2	3	1	3	4	6	1	5
3	4	6	6	3	4	2	1	1
4	5	6	3	4	5	1	3	5
4	2	6	1	2	5	3	6	4

Can you tell which outcome had the highest frequency? The lowest frequency?

If you rolled this dice another 36 times, would you expect to get the same results?

1 The table shows the results of a survey about the number of days of school students missed in a month.

Outcome	Tally	Frequency
0	ЖII	
1	ЖЖI	
2	Ж	
3	III	
4	II	
5 or more	II	

   a Complete the frequency column.

   b How many students are in this class?

   c What is the most common number of days missed?

   d How many students missed four or more days?

2 State whether each type of data is categorical, ordinal, discrete or continuous. Give reasons for your decisions.

   a year of birth

   b shoe size

   c type of fossil fuel

   d number of grains of sand on a beach

   e per cent of plastic that is recycled

   f length of a piece of wood

   g household waste produced per year.

3

   a Set up a frequency distribution table for the shapes in the diagram.

   b How many shapes are there?

   c Which shape appears most?

   d What percentage of the shapes are triangles?

 Fact

Some estimates put the number of stars in the known universe at between 5 and 10 times the number of grains of sand on all the world's beaches.

4   Mr Merlo gave his class a quiz of 10 maths questions. The number of
    questions each of the students solved correctly is listed below.

9	7	8	6	4	7	10	9	8	7
6	7	5	5	10	9	8	8	7	9
10	5	7	6	8	8	7	9	6	9

a   What was the highest score?

b   What was the lowest score?

c   Organise the data into a frequency distribution table.

d   What was the most common score?

e   How many students solved fewer than seven questions correctly?

f   What percentage of students had fewer than two incorrect answers?

9.3 Graphs

9.3.1 Frequency histograms and polygons

 Explore 9.4

Look back at the data for rolling a dice 36 times from Explore 9.3.
How would you represent this data to make it easy to read?

 Fact

When a histogram
represents qualitative
data, it is called a **bar
graph** or **bar chart**.

Another way of showing data in an easy-to-read form is to draw a graph.
Two commonly used types of graph are the frequency histogram and the
frequency polygon. The pencil data in Amina's survey in the previous section
can be graphed as a frequency histogram
or a frequency polygon.

A **frequency histogram** is a type of column
graph where the columns are drawn next to
each other.

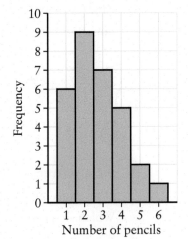

In a frequency histogram:

* each axis is labelled
* the height of each column is the frequency of that outcome
* the columns are centred on the outcomes

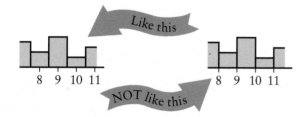

* the first column begins one-half of a column width in from the vertical axis.

A **frequency polygon** is a type of line graph. The axes are drawn in the same way as for a histogram.

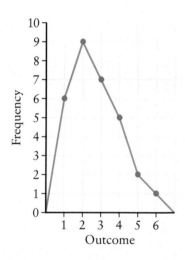

In a frequency polygon:

* each axis is labelled
* the dots showing the data are joined by straight lines
* the first and last dot are connected to the horizontal axis
* the first score is one unit in from the vertical axis.

If a frequency histogram and polygon are both drawn on the same axes, the polygon joins the midpoints of the top of each column in the histogram.

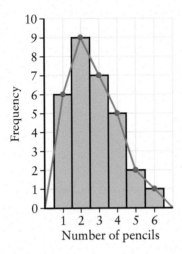

A **dual bar chart** or **composite bar chart** shows more than one set of data. It has a key to show what the different coloured bars represent.

🔍 Worked example 9.4

The dual bar chart shows the numbers of adults and children using a swimming pool on different days.

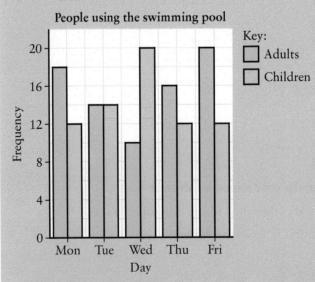

a   How many children used the pool on Wednesday?

b   How many adults used the pool on Wednesday?

c   How many children in total used the pool over the five days?

d   Which day had equal numbers of children and adults?

e   Which was the most popular day for adults?

Solution

a Draw a line from the top of the bar representing children on Wednesday to the frequency axis. Then read the value from the frequency axis.

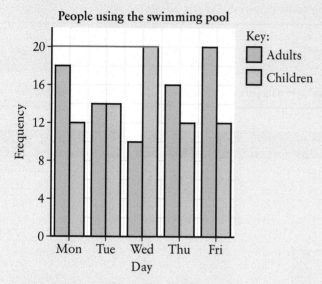

20 children used the pool on Wednesday.

b Draw a line from the top of the bar representing adults on Wednesday to the frequency axis. Then read the value from the frequency axis.

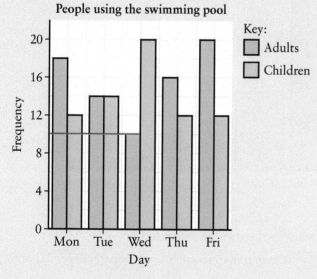

10 adults used the pool on Wednesday.

c Read across from the top of each children bar and add the totals.

12 + 14 + 20 + 12 + 12 = 70

70 children used the pool in total over the five days.

d The only day on which the two bars are the same height is Tuesday, so Tuesday had equal numbers of children and adults.

e We need to find the tallest adult bar. This is 20 adults, on Friday. Friday was the most popular day for adults.

Hint

This is the output of a GDC for the number of pencils brought to class.

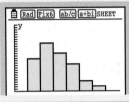

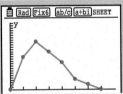

This is the output for the swimming pool data from Worked example 9.4.

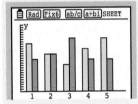

Reflect

Research how you can use software or your GDC to draw graphs.

Practice questions 9.3.1

1 The graph shows the average income per person per day in different countries.

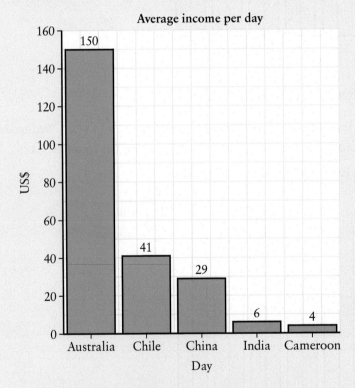

Average income per day

a What currency is the income measured in?

b What is the average daily income of someone living in China?

c What is the difference in the average daily income per person between Chile and Cameroon?

d How many times greater is the average daily income per person in Australia than in India?

2   This histogram shows the sizes of households in Grossau, a small village in northern Austria.

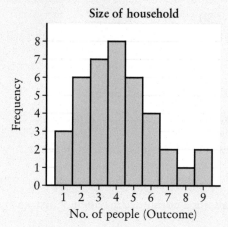

Size of household

a   What is the most common number of people per house?

b   How many people live by themselves?

c   What is the largest number of people in any one house?

d   How many houses have at least five people in them?

e   Work out the total number of houses in Grossau.

f   How many people live in groups of three?

g   Calculate how many people live in Grossau altogether.

3   The frequency polygon shows the marks Ms Coyne's MYP1 class achieved in a history test. The test was marked out of 60.

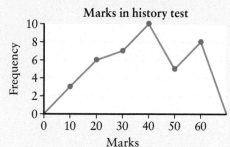

Marks in history test

a   Set up a frequency table by reading the frequency polygon.

b   How many students took the test?

c   How many students scored 20 or less?

4   The packaging on a bag of candies claims that there are 30 candies in each bag.

When the contents of 50 bags were counted, the following results were obtained.

31	30	29	30	30	31	32	29	28	30
30	29	30	31	33	32	31	30	28	29
27	31	30	32	30	30	31	30	32	31
30	29	28	30	31	28	30	30	27	30
33	29	32	30	30	31	30	31	32	30

a   Determine the highest and lowest numbers in the data.

b   Organise the data into a frequency distribution table.

c   How many packets did not have at least 30 candies in them?

d   How many packets had more than 30 candies?

e   Graph these results by drawing a combined frequency histogram and polygon.

5   Jacek was concerned about his country's gun laws. He asked 100 people to respond to the following statement:

> It should be made harder for people to own a gun.
> **A**   strongly agree   **B**   agree
> **C**   disagree   **D**   strongly disagree

These were his results.

A	B	A	A	C	A	A	D	A	B
C	B	D	A	B	B	C	B	A	B
B	C	B	B	A	B	C	C	B	B
D	B	A	A	B	C	C	B	B	B
B	B	A	B	C	B	B	C	A	B
B	A	A	B	B	C	A	A	A	A
B	A	C	D	A	B	B	A	C	B
B	B	B	C	A	B	B	C	B	D
B	A	B	B	B	C	C	C	A	B
B	C	C	B	B	B	A	A	C	A

a   Organise these results into a frequency distribution table.

b   How many people strongly agree with the statement?

c   What percentage of the people strongly disagree with the statement?

d   What percentage of the people asked think that it should be made harder for people to own a gun?

e   What percentage think that it should not be made harder?

f   Draw a frequency histogram of these results.

6   The dual bar chart shows the area of deforestation in Brazil and Southeast Asia over 15 years.

🏆 Challenge Q6

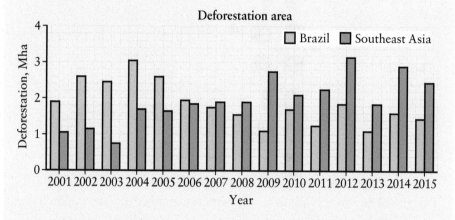

◎ Fact

Mha means mega hectare, a unit of measurement equal to 10 square kilometeres.

a   In which year was deforestation in Brazil first lower than in Southeast Asia?

b   What was the total area of deforestation in 2014?

c   What percentage of the area in part b can be attributed to Southeast Asia?

🔍 Investigation 9.1

Find out how to use a spreadsheet or other graph-drawing program to draw a bar chart.

Collect some data to investigate a question you are interested in.
For example:

- What is the most common type of vehicle to pass your house?
- What social media platforms do the students in your school prefer?
- How have cheetah populations in Africa changed over the last 20 years?

You could collect the data yourself, or use a library or the internet.

Use the software methods you learned to draw a bar chart to show your data.

Try using different scales. Which scale shows your data most clearly.

🎓 Research skills

Many energy companies provide their customers with bar charts showing how their energy consumption changes throughout the year.

9.3.2 Pictograms and pie charts

Explore 9.5

Day of week	Number of letters
Monday	● ◖
Tuesday	● ◗
Wednesday	● ● ◣
Thursday	● ●
Friday	● ◗
Saturday	● ●
Sunday	● ◖

This pictogram uses symbols to represent the number of letters in each word.

How many letters do you think a circle represents? How many letters does half a circle represent? Explain how you worked out what each symbol represents.

How would you represent the number of letters in your name?

A **pictogram** uses symbols to represent frequency. It has a key to show what each symbol represents.

This pictogram shows the recommended hours of sleep each night for different age groups.

Key: 🛏 represents 1 hour, 🛏 represents 30 minutes

Age	Recommendation
0 to 1 year	🛏🛏🛏🛏🛏 🛏🛏🛏🛏🛏 🛏🛏🛏🛏🛏 🛏
2 to 5 years	🛏🛏🛏🛏🛏 🛏🛏🛏🛏🛏 🛏🛏🛏

6 to 12 years	(bed symbols)
13 to 18 years	(bed symbols)
19 and older	(bed symbols)

A whole bed represents 1 hour and half a bed represents 30 minutes, or half an hour. In the 0 to 1 year age group, there are 15 whole beds and 1 half bed, so the recommended hours of sleep for this age group is 15.5. We can count the symbols for the other age groups and represent the same data in a frequency table.

Age	Hours
0 to 1 year	15.5
2 to 5 years	12.5
6 to 12 years	10
13 to 18 years	9
19 and older	8

Explore 9.6

These pie charts show how Jan and Marco spent their summer holiday. Blue represents the time spent at the seaside, green the time spent with parents and red the time they stayed at home.

Jan 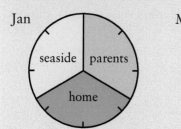   Marco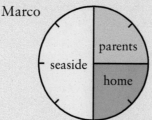

Can you compare how Jan and Marco spent their summer holiday?

A **pie chart** represents data in sectors, like 'slices' of a pie.

This pie chart shows the relative numbers of sea-turtle nests on one stretch of Florida beaches in 2020.

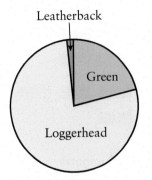

The sector for Leatherback turtles is by far the smallest. This shows that Leatherback turtle nests make up only a small percentage of all turtle nests on this stretch of beach.

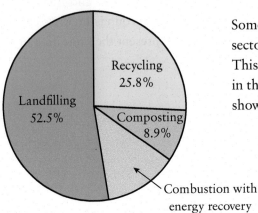

Sometimes the percentage for each sector is shown on the pie chart. This pie chart shows waste disposal in the USA in 2015. The largest sector shows landfilling, with 52.5%.

If the percentages are not shown, you might need to work them out yourself using the angles of the sectors.

◉ Worked example 9.5

The pie chart shows the waiting times for 40 patients at a doctor's surgery.

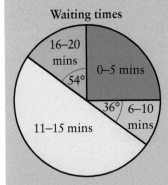

Find the number of people who waited:

a 0–5 minutes   b less than 11 minutes   c more than 15 minutes.

Solution

We know that the total number of patients is 40. We can use the angles in the pie chart to calculate the fraction of patients in each sector. Then we can find how many people that fraction represents.

a The sector labelled 0–5 minutes is $\frac{1}{4}$ of the pie chart.

$\frac{1}{4}$ of 40 = 10

10 people waited 0–5 minutes.

b We need to add the number of people in the 0–5 minutes, and 6–10 minutes, sectors.

A full circle is 360°. The 6–10 minutes, sector is 36°. We need to find what fraction of 360° this is.

$\frac{36}{360} = \frac{1}{10}$, so 36° is $\frac{1}{10}$ of 360°

$\frac{1}{10}$ of 40 = 4

4 people waited 6–10 minutes.

10 + 4 = 14

14 people waited less than 11 minutes.

b The 16–20 minutes sector is 54°.

$\frac{54}{360} \times 40 = 6$

6 people waited more than 15 minutes.

Worked example 9.6

The frequency table shows the colours of students' mobile phone covers.

Colour	Frequency
White	7
Silver	8
Black	5
Blue	1
Pink	3

Draw a pie chart to represent this data.

Solution

There are 24 students in total. The angle to represent one student is $\frac{360}{24} = 15°$

Multiply 15° by the frequency for each colour to find the angle for that colour.

Colour	Frequency	Sector angle
White	7	7 × 15° = 105°
Silver	8	8 × 15° = 120°
Black	5	5 × 15° = 75°
Blue	1	1 × 15° = 15°
Pink	3	3 × 15° = 45°
Total	24	360°

Add a column to calculate the angle of each sector.

Add a row for the totals. The total frequency is 24.

Check that the angles add to 360°.

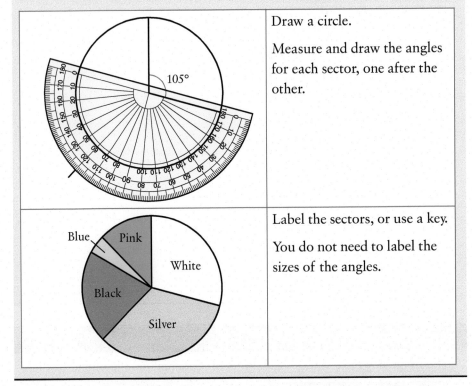

Draw a circle.

Measure and draw the angles for each sector, one after the other.

Label the sectors, or use a key.

You do not need to label the sizes of the angles.

Can you see anything wrong with this pie chart? What information is missing?

Reflect

How does adding rows and columns to the frequency table help you to organise your work?

When a pie chart shows percentages, what size of angle represents 1%?

1   This symbol represents ten pineapples.
    What do these symbols represent?

    a

    b

    c

    d

2   This pictogram shows the numbers of sandwiches sold at Harriet's
    Café one day.

    Sandwiches sold

Egg	⊠ ▽
Salmon	⊠ ⊠ ⊠
Chicken	⊠ ⊠ ▽

    Key ⊠ represents 4 sandwiches

    a   How many salmon sandwiches were sold?

    b   How many more chicken sandwiches were sold than egg
        sandwiches?

    c   Nine cheese sandwiches were sold. Draw symbols to represent this.

3   40 people were asked: 'What is
    your favourite type of food?'
    The table shows the results.

Type of food	Frequency
Italian	9
Chinese	10
Indian	8
Thai	7
Mexican	4
Canadian	2

    a   Draw a bar chart and a pie
        chart for this data.

    b   If you want to compare the
        results for types of food,
        what is a better tool to use,
        bar or pie chart?
        Give a reason for your answer.

4   The pie chart shows coal exports from different countries.

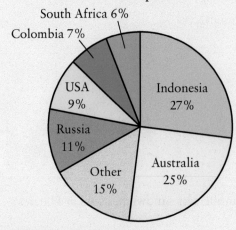

Coal exports

South Africa 6%
Colombia 7%
USA 9%
Russia 11%
Other 15%
Indonesia 27%
Australia 25%

The total amount of coal exported is 1137 megatonnes.
How many megatonnes of coal are exported by the 'Other' countries?

5   This pie chart shows the numbers of people in each car entering a car park.

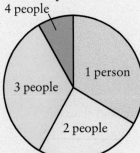

People in cars

4 people
1 person
3 people
2 people

a   Measure the angle of the sector for cars with only one person.
    What fraction of the cars have only one person?

b   The pie chart represents 60 cars in total.
    How many of these cars had four people in them?

6   Lucca carried out a survey to ask people's
    favourite type of film.
    The pie chart shows the results.

Film choices

Romance
Comedy
Other
Action

a   16 people chose Comedy.
    How many people chose Action?

b   How many people's answers are
    represented in the pie chart?

7   The pie chart shows the cost distribution of a $2.40 chocolate bar.

Chocolate cost distribution

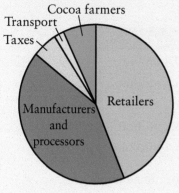

Department	Share
Processors	22 cents
Transport	8 cents
Cocoa farmers	26 cents
Taxes	12 cents
Manufacturers	78 cents
Retailers	94 cents

a   Write down the central angle corresponding to each sector.

b   Who gets the largest share of the price of each chocolate bar?

c   What percentage of the price goes to the cocoa farmers?

8   36 students were asked, 'How do you find out about news in your community?'

The table shows the results.

Subject	Frequency
Social media	18
Website	9
Bulletin board	6
Word of mouth	3

a   Work out the size of the angle to represent one student.

b   Draw a pie chart to show these results.

9   A group of students were asked which device they used most to access social media.

The table shows the results.

Device	Frequency
Laptop	3
Mobile phone	11
Tablet	4
None	2

a   Draw a pie chart for this data.

b   What percentage of students use mobile phones the most?

Hint Q8

Label the sectors, or make a key. Give your chart a title.

10 In 2016, it was reported that across the world, 152 million children were in child labour. The regional prevalence of child labour is given in the table.

Region	Number
Africa	30.8 million
Americas	8.1 million
Asia-Pacific	11.2 million
Europe and Central Asia	6.2 million
All other countries	65.1 million

a   Draw a pie chart to represent this information

b   What percentage of child labour is in each region?

11 71% of the children in child labour work in agriculture, 12% work in industry, and 17% work in services.

Draw a pie chart to represent this.

Hint Q12

You will need to round the angles to the nearest degree. Check that they add up to 360°.

12 Cobalt is a valuable element used in mobile phone batteries.

The table shows the percentage of global cobalt supply mined in different regions.

Region	Percentage
Democratic Republic of Congo	61%
Asia-Pacific	16%
Africa	8%
Europe	6%
Latin America	5%
USA & Canada	4%

Draw a pie chart to represent this data.

13 These two pie charts show the proportions of businesses paying a low, medium or high wage in China and in Europe.

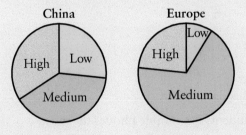

a   For each statement, write 'True', 'False', or 'Cannot tell from the chart'.

   i   The fraction of businesses paying high wages is smaller in China than in Europe.

   ii   The percentage of businesses paying medium wages is greater in Europe than in China.

   iii   The number of businesses paying medium wages is greater in Europe than in China.

b   What extra information would you need to compare the numbers of businesses paying low wages in China and in Europe?

9.4   Stem-and-leaf plots

 Explore 9.7

The histogram shows the number of books owned by students in a class.

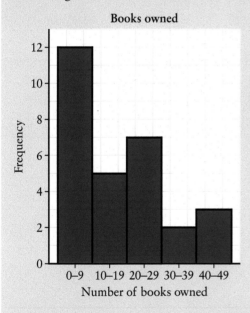

Books owned

Number of books owned

Can you use the histogram to find out how many children own fewer than 20 books? If so, explain how. If not, explain why not.

Can you use the histogram to find out how many students own 17 books? If so, explain how. If not, explain why not.

A **stem-and-leaf plot** resembles a histogram on its side. It shows the same information as a histogram, but the values of the individual data items are not lost.

Stem-and-leaf plots are used when the data set is not very large. Each leaf needs to be a single digit, so decimals may need to be rounded.

Worked example 9.7

Draw a stem-and-leaf plot for the following marks in a Spanish test.

68	59	85	53	57	93	84	73	41	65
77	73	66	50	97	67	62	54	64	88
48	80	68	66	71	91	79	73	84	75

Solution

We need to split each number into a **stem** and a **leaf**. In this case, the stem is the tens digit and the leaf is the ones digit.

Stem	Leaf
4	1 8
5	9 3 7 0 4
6	8 5 6 7 2 4 8 6
7	3 7 3 1 9 3 5
8	5 4 8 0 4
9	3 7 1

The smallest data item is 41, and the largest data item is 97, so the stems go from 4 to 9. We write these in ascending order in the left column.

We go through each data item in turn and write its ones digit in the row corresponding to its tens digit. The first data item is 68. The tens digit is 6, so we write an 8 in the '6' row.

To make an ordered plot, we need to rewrite the leaves in ascending order and write a key.

Spanish test marks:

Key: 4 | 1 = 41 marks

Stem	Leaf
4	1 8
5	0 3 4 7 9
6	2 4 5 6 6 7 8 8
7	1 3 3 3 5 7 9
8	0 4 4 5 8
9	1 3 7

A **back-to-back stem-and-leaf plot** allows us to compare two groups of data at a glance.

Back-to-back stem-and-leaf plots are useful for comparing two similar sets of data, such as marathon finishing times for different groups of people.

Reflect

Class MYP1 section A and section B were given a statistics test.
The results are shown on the back-to-back stem-and-leaf plot below.
Marks are out of 100.

Marks on statistics test:

Key: For Section A, 3 | 6 = 63. For Section B, 5 | 1 = 51

Section A	Stem	Section B
9 8 6 6 0	3	
9 7 7 3 1 1 1	4	3 6 6 8
8 8 5 3 3 0	5	1 7 9 9
9 8 5 7 3	6	3 8 9
	7	0 5 5 5 8 9
	8	2 6 7 7
0	9	0 0 1 3

Explain how the back-to-back stem-and-leaf plot works.

Compare the performance of the two sections.

Practice questions 9.4

1 Prepare a stem-and-leaf plot for each set of test marks. Make sure to add a key.

a

26	15	62	51	47	19	33	40	58
26	38	35	47	41	36	29	17	16
62	28	18	47	30	26			

b

53	81	74	72	91	45	80	57	44
82	70	67	59	80	81	60	49	77
65	42	73	63	92	88			

2 The number of minutes it takes 22 students to get to school are recorded below as two-digit numbers. The key is: 1 | 5 = 15 minutes

Stem	Leaf
0	8 6 4 1 9 2
1	5 8 9 1
2	0 7 6 9
3	6 3 2
4	6 5 0 3
5	
6	2

a Rewrite the stem-and-leaf plot as an ordered plot.

b Construct a frequency table for the stems.

c What percentage of the students take less than 20 minutes to get to school?

Hint Q2b

The row heading for the 1 stem will be 10–19.

3 Two types of batteries used in modern phones are lithium polymer and lithium ion. The times, in hours, that different phones stayed charged were recorded and the results are shown below.

Lithium polymer

88	55	60	70	40	48	36	80	74	68	56	78	82	38	46	60
60	82	94	40	70	56	90	44	66	52	84	50	60	42	84	38

Lithium ion

36	64	40	48	22	70	62	84	18	76	54	32	52	32
80	46	74	62	52	60	84	42	36	82	84	44	50	64

a Prepare a back-to-back stem-and-leaf plot for this data.

b Is there a difference in performance between the two types of battery? Give a reason for your answer.

Challenge 4

4 The pulse rates of 24 adults and 24 children were measured and recorded. The results are shown below.

Adults

12	64	59	60	72	61	58	55	76	76	80	81
58	64	72	60	92	56	57	68	80	68	72	64

Children

102	86	67	85	98	77	81	85	72	62	88	92
86	82	94	78	100	94	85	95	99	86	77	99

a Use a back-to-back stem-and-leaf plot to organise this data.

b Describe each set of scores and compare them.

9.5 Scatter plots

A **scatter plot** (also called a **scatter diagram** or **scatter graph**) uses dots to represent values for two different quantitative variables. Each dot represents two values: one for the variable on the horizontal axis and one for the variable on the vertical axis.

Scatter plots are used to identify relationships between two variables. The pattern created by the dots tells a story. Look at the scatter plot on the next page. The dots that have a high value for weight tend to have a high value for height too. It appears that as weight increases, height also tends to increase.

If one variable increases as the other variable increases, we say that the variables have a **positive relationship**.

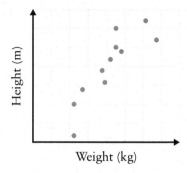

If one variable decreases as the other variable increases, we say that the variables have a **negative relationship**.

If the points are spread all over the scatter plot with no obvious pattern, we say that the variables have **no relationship**.

◈ Fact

Is there always a relationship?

A scatter plot of date of birth and weight of individuals shows that there is no connection between those two variables.

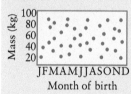

Worked example 9.8

The scatter plot shows the mid-year exam scores for 20 students in science and maths. Both exams had 10 possible marks. The dots for two of the students are labelled.

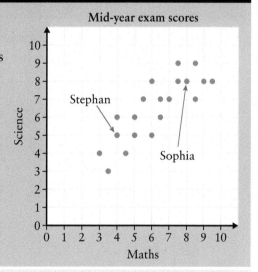

a Is there a relationship between exam scores for the two subjects? Give a reason for your answer.

b What scores did Sophia and Stephan get in each subject?

Solution

a We need to look at the shape formed by the data points.

The points roughly follow a line from bottom left to top right. This shows that maths and science are, in general, positively related. Low maths marks are associated with low science marks, and high maths marks are associated with high science marks.

b Sophia got 8 marks in both maths and science. Stephan got 4 marks in maths and 5 marks in science.

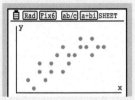
Investigation 9.2

Collect quantitative data about the students in your class. Some examples of data you might collect are:

- age in months
- height
- number of siblings
- time spent reading each week
- time spent playing sport each week.

Choose two of the data sets you have collected and draw a scatter plot to represent them. Is there a relationship between the two data sets? If there is, describe it.

Try another two data sets. Can you find a pair of data sets with no relationship? Can you find a pair of data sets with a negative relationship? What do these relationships look like on a scatter plot?

Practice questions 9.5

1 For each of the scatter diagrams below, state whether there is a positive relationship, a negative relationship or no relationship.

a **Running and maths**

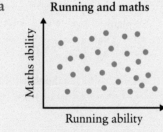

b **Bird seed costs**

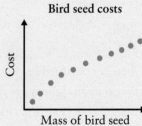

c **Smoking and death**

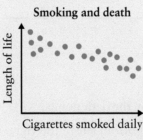

d **Television and study**

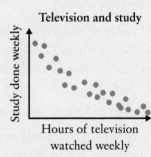

e **Bowling and arm length**

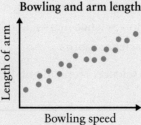
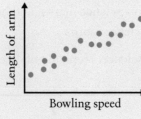

2 The scatter plot shows maths and English marks for the students in a class.

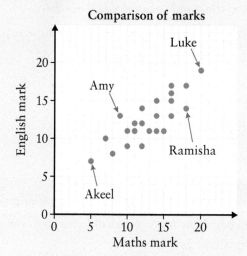

Comparison of marks

a Write the maths and English marks scored by:

  i Akeel   **ii** Luke   **iii** Amy   **iv** Ramisha.

b What was the highest:

  i maths mark   **ii** English mark?

c What relationship is suggested by this scatter plot?

3 Mr Pearson wanted to see if his students achieved similar marks in both the mid-year and end-of-year exams. The table shows the results from both tests.

Hint Q3

Use a spreadsheet or GDC to plot the data.

Mid-year	End-of-year	Mid-year	End-of-year
27	45	48	54
69	72	63	69
45	42	60	66
66	69	69	63
96	90	84	69
93	90	96	87
90	78	87	72
54	57	75	75
84	78	75	81
75	63	30	36
39	48	57	51
93	84	78	72

a Draw a scatter plot to show these results.

b Is there any relationship between the tests? Give a reason for your answer.

? Check your knowledge questions

1 A group of people were asked, 'How many times did you go to the gym last week?' The results are shown below.

3	2	1	0	6	1	1	0	4	0
7	0	1	3	5	6	1	2	1	5

a Copy and complete the tally chart for the data.

Number of gym visits	Tally	Frequency
0		
1		

b Draw a bar chart for this data.

c How many people answered the question?

d What conclusions, if any, could you draw from these data?

2 Alesha writes this question for a survey.

How many times did you exercise last week?

1–2 times	☐
3–4 times	☐
4–5 times	☐
more than 5 times	☐

a Explain two things that are wrong with this question.

b Write a better version of this question.

3 This pictogram shows the drinks bought from a drinks machine one day.

Key: ⬜ = 10 drinks

Tea	
Coffee	
Hot chocolate	

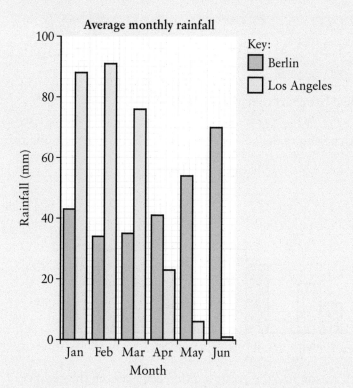

a How many drinks were bought in total?

b How many more coffees were sold than hot chocolates?

c What fraction of the drinks sold were tea?

4 The bar chart shows the average monthly rainfall in Berlin and Los Angeles during the first six months of the year.

a In which months does Berlin have less rainfall than Los Angeles?

b Which is the driest month in:
   i Berlin   ii Los Angeles?

c   Which month has the smallest difference in rainfall between the two cities?

5   Cherry asked the students in her class what they spend most of their money on. The results are shown in the table.

Most money spent on	Frequency
Music	3
Food and drink	10
Clothes	5
Phone contract	12
Other	6

Draw a pie chart to represent this information.

6   A chicken farmer sorts the chickens' eggs into small, medium and large.

The table shows the numbers of eggs of different sizes from two breeds of chicken, the Rhode Island Red and the Moran.

	Small	Medium	Large
Rhode Island Red	16	40	24
Moran	8	32	20

Draw a dual bar chart to represent this data.

 Challenge Q7

7   The bar chart shows the number of visitors to a theme park one year.

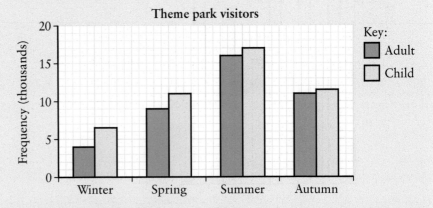

a   Which season had the greatest difference between the number of adults and the number of children?

b   How many child visitors were there in the summer?

c   How many visitors were there in total in the autumn?

d   The park manager says, 'There were more visitors in the summer than in the winter and spring combined.' Show that the manager is correct.

8   The table shows the amount of water used per person each day in different countries, rounded to the nearest 100 litres.

Challenge Q8

Country	Water used
USA	1200
Canada	900
Mexico	700
Korea	500
Brazil	400
France	400
Germany	300

Draw a pictogram to represent this data.

9   Find bar charts or pictograms on a topic that interests you. You can search online, in newspapers, books or magazines. Write some questions for your classmates to answer about these charts.

Challenge Q9

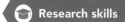
Research skills

10 The tables show the performance of students in art and music.

a   Draw a scatter plot to represent this information.

Art	60	50	55	70	75	56	60	80	85	40	56
Music	80	90	85	70	65	52	60	60	55	50	58

Art	52	73	45	90	68	65	50	80	70	65	
Music	59	70	49	50	72	75	50	40	50	60	

b   Is there a relationship between art marks and music marks? Give a reason for your answer.

Chapter 1 answers

Do you recall?

1 Directed numbers are integers of a particular size and direction (positive or negative).

2 Number operations are addition, subtraction, multiplication and division.

3 Student's own answer, for example, partitioning

4 Student's own answer, for example, rounding and then adjusting

Practice questions 1.1

1 a 6003   b 9000

   c 455   d 12449

2 a 9232   b −5486   c 10760

   d −759   e −1   f 13489

3 94

4 −35

5 a 53, 21, 14451

   b −7, 53, 21, −2, 14451

   c −7, 0.4, 53, $\frac{4}{5}$, 34.5, 21, −2, −4.75, 14451, $\frac{423}{1000}$

6 a False   b True

   c True   d False

   e True   f True

   g True   h False

7 2

8 a 14, 15, N   b −13, −11, Z

   c 0.9, 1.1, R   d $\frac{15}{5}, \frac{18}{5}$, R

   e 10, 15, Z   f 12600, 12650, N

9 a Any fraction or decimal between 1 and 2, e.g. $1\frac{1}{2}$, 1.5, 1.4

   b Any fraction or decimal between $\frac{2}{5}$ and $\frac{3}{5}$, e.g. $\frac{1}{2}$, 0.5, $\frac{5}{10}$

   c Any fraction or decimal between 21.7 and 21.8, e.g. 21.75, 21.77

   d Any fraction or decimal between −49 and −48, e.g. −48.9, −48.5

   e Any fraction or decimal between 0.1 and 0.15, e.g. 0.11, 0.14

   f Any fraction or decimal between −5.5 and −5.6, e.g. 5.53

10 a Student's own set of five natural numbers

   b Student's own set of five integers

   c Student's own set of five real numbers

11 a 0, 1, 2, 3, 4, 5, 6, 7, 8, 9, 10, 11, 12, 13, 14, 15, 16, 17, 18, 19

   b 1 (0 + 1), 3 (1 + 2), 5 (2 + 3), 7 (3 + 4), 9 (4 + 5), 11 (5 + 6), 13 (6 + 7), 15 (7 + 8), 17 (8 + 9), 19 (9 + 10)

12 5050, (1 + 100) × 50, see Gauss' method

Practice questions 1.2.1

1 a 1240   b 107

   c 1005   d 10436

   e 6351   f 23508

   g 67137   h 701083

   i 950862   j 57006

   k 1312   l 3221

   m 4116

2 a 2735   b 412917

   c 302109   d 1000024

   e 5841289   f 56006770

   g 923441711   h 308005049

3 a Seven thousand, nine hundred and twenty-four

   b Forty-one thousand and fifty-four

   c Fifty-seven thousand, two hundred and two

   d Five hundred and forty-seven thousand, one hundred and twenty-six

   e Nine hundred and eight thousand and forty

   f Eight million, five hundred and forty-five thousand, seven hundred and eighty-two g Fifty-nine million, twenty-five thousand, one hundred and eighteen

h Twenty million, one hundred and five thousand and sixty seven

4 a 3, 40 and 500

b 3000, 40 and 5

c 30 000, 400 and 5

d 3000, 4 and 50 000

e 300, 400 000 and 5

f 3 000 000, 400 000 and 50 000

g 300 000, 40 000 000 and 50

h 300 000 000, 4000 and 50 000 000

5 a 600, 50 and 8

b 9000, 200, 50 and 4

c 60 000, 3000, 100, 20 and 5

d 600 000, 20 000 and 400

e 900 000, 200, 10 and 8

f 5 000 000, 400 000, 70 000, 800, 10 and 6

g 50 000 000, 9 000 000, 20 000, 5000, 600, 10 and 8

h 600 000 000, 70 000 000, 800 000, 9000, 200, 40 and 3

6 a 28 348

b 732 816

c 5 812 973

d 9 067 049

e 9 607 490

f 5 081 297

g 61 080 937

h 9 006 503

7 a 40 000 + 9000 + 500 + 20 + 3

b 50 000 + 4000 + 800 + 70

c 800 000 + 70 000 + 200 + 50 + 2

d 7 000 000 + 500 000 + 20 000 + 1000 + 400 + 60 + 3

8 a 83 658, 84 125, 84 874, 85 152, 85 252

b 45 000, 45 004, 45 025, 45 688, 45 842

c 749 254, 749 478, 749 564, 749 887, 752 054

d 1 235 465, 1 235 589, 1 235 854, 1 238 008, 1 259 471

e 1 036 987, 1 775 258, 1 897 026, 1 897 278, 1 897 429

f 87 981, 88 123, 88 598, 88 742, 89 056

g 100 000, 100 054, 100 471, 101 225, 101 471, 101 568

h 6 126 700, 6 127 554, 6 541 965, 6 600 000

Practice questions 1.2.2

1 a 7 tenths or $\frac{7}{10}$

b 7 hundredths or $\frac{7}{100}$

c 7 thousandths or $\frac{7}{1000}$

d 7 ones

e 7 hundredths or $\frac{7}{100}$

f 7 thousandths or $\frac{7}{1000}$

2 a $3, \frac{4}{10}$ b $3, \frac{4}{100}$

c $\frac{3}{10}, \frac{4}{100}$ d $\frac{3}{100}, \frac{4}{1000}$

e $\frac{3}{1000}, \frac{4}{10}$ f $40, \frac{4}{100}, \frac{3}{1000}$

3 a 0.472   b 3.926

c 0.5941   d 5.0095

e 0.7016   f 9.07002

g 45.908   h 67.04

4 a 0.7 + 0.08 + 0.005

b 2 + 0.5 + 0.06

c 0.7 + 0.02 + 0.005 + 0.0003

d 5 + 0.08 + 0.007

e 0.6 + 0.004 + 0.0001

f 8 + 0.5 + 0.002 + 0.0006

g 20 + 4 + 0.8 + 0.001 + 0.0003

h 10 + 3 + 0.02 + 0.0005

5   a   True   b   False

    c   True   d   False

    e   False   f   True

    g   False

6   a   0.3, 0.4, 0.7, 0.8

    b   1.9, 3.11, 3.24, 9.04

    c   0.3, 0.33, 0.7, 0.71, 1

    d   0.06, 0.6, 6, 6.06, 6.6

    e   0.04, 0.4, 0.44, 4.04, 4.4

    f   0.4, 0.45, 0.49, 0.5, 0.51

    g   0.684, 0.75, 0.758, 0.8, 0.81

    h   0.09, 0.1, 0.101, 0.8, 0.87

7   a   Any number with the digit 3 in the tens place, e.g. 30.124 or 34.102

    b   Any number with the digit 3 in the thousandths place, e.g. 12.043 or 40.123

    c   43.201 or 43.210

    d   Any number with the tens digit less than 4, e.g. 34.102 or 10.243

Practice questions 1.3.1

1   a   71   b   515

    c   71   d   57

    e   193   f   264

    g   4234   h   1269

    i   306   j   1618

2

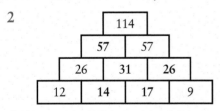

3   a   2   b   −9   c   0   d   −4

    e   12   f   −8   g   11   h   −65

4   a   −3   b   −5   c   −6   d   −10

    e   −10   f   −19   g   −19   h   −39

    i   −77   j   −200

5   a   7   b   18   c   20   d   10

    e   0   f   −3   g   11   h   4

    i   11   j   −300

6   a   87   b   54   c   31   d   −19

    e   −7   f   −55   g   19   h   148

    i   −45   j   517

7

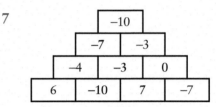

8   a

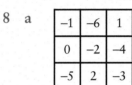

    Or, any rotation or reflection of this solution.

    b   −6

9   126

10   No, for example, 277 − 53 = 224, which is less than the required 230

11   62

12   36

13   −8 − (−27) is the same as −8 + 27 = 19

14   353 + 247 = 600

15   a

$$
\begin{array}{ccc}
 & 3 & \boxed{6} & 7 \\
+ & \boxed{5} & 4 & 8 \\
\hline
 & 9 & 1 & \boxed{5} \\
\end{array}
$$

    b

$$
\begin{array}{ccc}
 & 6 & 4 & \boxed{7} \\
- & 2 & \boxed{1} & 8 \\
\hline
 & \boxed{4} & 2 & 9 \\
\end{array}
$$

16 a $1 + 9 = 3 - (-7)$ or any answer from part b, below

 b $(-7) + 1 = 3 - 9$, $9 + (-7) = 3 - 1$

17 −13 and 12

18 27 fewer orangutans than the target of 200 were released or the target of 200 was missed by 27.

Practice questions 1.3.2

1 a 5   b 4   c 8   d 6
   e 9   f 9   g 6   h 9
   i 2   j 7   k 8   l 5
   m 7   n 7   o 9   p 9
   q 6   r 5   s 9   t 9
   u 6

2

×	3	6	0	1	9	5	10	7	2	8	12	4
4	12	24	0	4	36	20	40	28	8	32	48	16
6	18	36	0	6	54	30	60	42	12	48	72	24
7	21	42	0	7	63	35	70	49	14	56	84	28
8	24	48	0	8	72	40	80	56	16	64	96	32

3 Students show a written method for each calculation, for example, long multiplication or the Japanese method.

4 a $2\frac{2}{5}$ b $2\frac{3}{4}$ c $3\frac{3}{10}$ d $5\frac{5}{8}$
   e $1\frac{5}{7}$ f $1\frac{1}{4}$ g $3\frac{4}{5}$ h $10\frac{2}{3}$
   i $2\frac{3}{4}$ j $11\frac{3}{5}$ k $18\frac{3}{4}$ l $11\frac{9}{10}$

5 a $301\frac{9}{10}$ b 2500   c $1257\frac{7}{10}$
   d $149\frac{1}{4}$ e $211\frac{4}{5}$ f 121
   g 608   h 6008   i $1571\frac{5}{8}$

6 a $6\frac{1}{4}$ b 6   c 8
   d 9   e $11\frac{12}{17}$ f $11\frac{7}{18}$
   g $13\frac{1}{3}$; accept any equivalent fraction

 h $23\frac{7}{23}$

 i $53\frac{9}{16}$

7 23 minibuses

8 $18\frac{2}{3}$ days

9 Students show a written method for each calculation, for example, long multiplication or the Japanese method.

10 $3901, the total of 22 × $170 = $3740 and 7 × $23 = $161

11

		5	4	9
×				8
4	3	9	2	

12 a $175\frac{1}{5}$ b $70\frac{7}{8}$

13 a 43 180   b 431 800
   c 34   d 8636
   e 1270   f 68
   g 2159   h 508

Practice questions 1.4.1

1 a 46   b 79   c 9   d 2
   e 10   f 4   g 4   h −2
   i 11   j −1

2 a 200   b 36   c 80   d 60
   e 2   f 2   g 3   h 8
   i 8   j 15

3 a 4   b 22   c 2   d 12
   e 1   f 58   g 55   h 0

4 a 21   b 45   c 64   d 15

5 a =   b ≠   c ≠

6 a $3 + 4 \times 5 = 23$
   b $4 \times 7 - 12 \div 4 = 25$
   c $6 - 30 \div 3 + 6 \times 7 = 38$
   d $8 + 12 \times 4 - 12 \div 2 = 50$

Practice questions 1.4.2

1. a 90 b 5 c 35

   d 3 e 855 f 4

   g 2 h 102 i 73

2. a 88 b 48 c 6

   d 4 e 60 f 14

   g 52 h 1 i 54

3. a 40 b 81 c 89 d 2

   e 26 f 14 g 184 h 10

4. a 4 b 5 c 4

   d 6 e 1 f 83

5. a Not needed as the 4 × 7 has priority over the +

   b Needed as the × would have priority over the + otherwise

   c Not needed as the × and ÷ already have priority over the +

   d Not needed as the × and ÷ already have priority over the +

   e Needed as the × and ÷ would have priority over the + otherwise

   f Needed as the × would have priority over the − otherwise

6. a (3 + 5) × 5 = 40

   b 4 × (28 − 10) ÷ 3 = 24

   c (20 − 6) × (5 + 6) ÷ 7 = 22

   d (5 + 2) × (10 − 2) ÷ 2 = 28

7. a (7 + 8) × 2 = 30

   b 4 × (4 + 7) = 44

   c 30 ÷ (3 + 2) = 6

   d (4 + 2) × (3 + 7) = 60

   e 4 + 2 × 3 + 7 = 17 Brackets are not needed but could be placed 4 + (2 × 3) + 7 = 17

   f (4 + 2) × 3 + 7 = 25

   g 100 − (5 + 3) × 6 = 52

   h 75 ÷ (5 ÷ 5) = 75

8. $\dfrac{(9 - 5) \times 3}{6} = 2$

9. There is more than one solution for many of these, but here are some examples:

   1 = (2 − 1) × (4 − 3)

   2 = 4 − 3 + 2 − 1

   3 = (4 − 1) × (3 − 2)

   4 = 4 + 3 − 2 − 1

   5 = (3 − 2) × (4 + 1)

   6 = 4 + 3 − 2 + 1

   7 = (4 + 3) × (2 − 1)

   8 = (4 − 2) × (3 + 1)

   9 = 3 × 2 + 4 − 1

   10 = 1 + 2 + 3 + 4

   11 = 1 × (2 × 4 + 3)

   12 = 1 + 3 + 2 × 4

   13 = 3 + 2 × (1 + 4)

   14 = 1 × (2 + 4 × 3)

   15 = 1 + 2 + 3 × 4

   16 = 2 × (1 + 3 + 4)

   17 = 2 + 3 × (4 + 1)

   18 = 1 × 3 × (2 + 4)

   19 = (2 + 3) × 4 − 1

   20 = 1 × 4 × (2 + 3)

   21 = 1 + (2 + 3) × 4

   22 = 2 × (4 × 3 − 1)

   23 = 2 × 3 × 4 − 1

   24 = 1 × 2 × 3 × 4

   25 = 1 + 2 × 3 × 4

Check your knowledge questions

1. 311

2. −17

3. a 22

   b −10, −3, 22

   c −10, −3, −2.5, $\frac{2}{3}$, 0.5, 22

4. a Student's own answer between $\frac{1}{3}$ and $\frac{2}{3}$, e.g. $\frac{2}{5}$, 0.5, $\frac{1}{2}$, 0.6

   b Student's own answer between −16 and −15, e.g. −15.5

c Student's own answer between 0.3 and 0.35, e.g. 0.31, 0.32, 0.302

5 a 1027   b 638 496

   c 504 201   d 3 000 008

6 a Three thousand, eight hundred and seventy-six

   b Fifty-two thousand, three hundred and seven

   c Four hundred and eighty-seven thousand and thirty-one

   d Sixty-five million, eight hundred and ninety thousand, three hundred and fourteen

7 a 2000 and 600

   b 20 000 and 6

   c 2 000 000 and 600 000

   d 2 tenths or $\frac{2}{10}$ and 6 thousandths or $\frac{6}{1000}$

   e 20 and 6 hundredths or $\frac{6}{100}$

   f 2 thousandths or $\frac{2}{1000}$ and 6 hundredths or $\frac{6}{100}$

8 a 2348, 23 476, 230 687, 2 376 400

   b 0.3, 0.312, 0.321, 0.33

9 a 3146   b 577

10 a 6   b −8   c −21

   d −28   e −32   f −8

11 Students show a written method, for example, long multiplication or the Japanese method.

12 a $5\frac{5}{7}$ b $6\frac{2}{3}$ c $6\frac{2}{8}$ or $6\frac{1}{4}$

13 a $629\frac{5}{8}$ b $2181\frac{9}{11}$

14 a $25\frac{1}{26}$ b $27\frac{1}{32}$

15 a 194 040   b 9702   c 231

   d 194 040   e 8400

16 a 16   b −8   c 7

17 a 360   b 5   c 16

18 a 26   b 18   c 12

   d 0   e 33

19 a 99   b 3   c 6   d 124

20 a 88   b 5   c 52

21 a 276   b 4

22 a 6   b 8   c 14

23 a Not needed as × and ÷ both take priority over +

   b Needed as without the brackets the × and ÷ would otherwise take priority over +

   c Needed as without the brackets the × would otherwise take priority over both −

24 $(7 \times 6) - 5 = 37$

Chapter 2 answers

Practice questions 2.1.1

1 130 pages

2 £22.55

3 13 cars

4 30 miles

5 6 slices

6 3520

Practice questions 2.2.1

1 José is 13.

2 12 and 18

3 5 and 7

4 11 chickens

5 6 cupcakes

6 £8.60

7 90 cents

8 20 yo-yos

9 4 Zigs and 7 Zags or 11 Zigs and 4 Zags or 18 Zigs and 1 Zag

10 Michael is 7 and Andre is 12.

Practice questions 2.2.2

1 4 ways (LPB, LBP, PLB, BLP)

2 HHH, HHT, HTH, HTT, THH, THT, TTH, TTT

3 10 different pairs (AB, AC, AD, AE, BC, BD, BE, CD, CE, DE)

4 19 numbers (3, 13, 23, 30, 31, 32, 33, 34, 35, 36, 37, 38, 39, 43, 53, 63, 73, 83, 93)

5 12 numbers (34, 35, 36, 43, 45, 46, 53, 54, 56, 63, 64, 65)

6 6 different ways (123, 132, 213, 231, 312, 321)

7 12 ways (Class A: MJ, MA, JM, JA, AM, AJ. Class B: CP, PC. 6 ways × 2 ways = 12 ways)

Practice questions 2.2.3

1 83 students

2 26 children

3 a 5 highlighters   b €1.50
   c €8.35   d €13.75

4

A question (10 points)	B question (5 points)	C question (2 points)	Total
1	1	7	29
1	3	2	29
2	1	2	29
0	1	12	29
0	3	7	29
0	5	2	29

5 6 carpenters go home by bus.

Practice questions 2.2.4

1 The number is 7.

2 Cuddles is a hippo.

3 The number is 25.

4 Olivia's aunt is 18.

5 18 cards

6 $4.80

7 Mila eats cereal, William eats fruit and Noah eats eggs.

8 Maura is the skeleton, Brigid is the witch, Tom is the pirate and Tristam is the rabbit.

9 Cathal Miller, Ryan Burns and Nicola Jones

10 Today is Friday.

11 Darren owns a dog, Cheng owns a turtle, Luke owns a goldfish and Ali owns a cat.

12 Sasha scored 11 points, Marta scored 23 points, Zoe scored 19 points, Caterina scored 25 points and Lynn scored 17 points.

Practice questions 2.2.5

1 The blue block

2 7.5 cm

3 12 posts

4 Friday

5 8 days

6 10 paths

7 15 handshakes

8 5th bounce

9 12 students

Practice questions 2.2.6

1 a 67   b 35   c $\frac{1}{2}$ d 7

2 a €14   b €24   c €44

3 a i 9 coins   ii 19 coins
      iii 99 coins

   b i 22 blocks   ii 34 blocks
      iii 298 blocks

   c i 31 pencils   ii 51 pencils
      iii 151 pencils

4 a 19 m b 38 m c 95 m d 950 m

5 a i 1 handshake   ii 3 handshakes
      iii 6 handshakes   iv 10 handshakes
      v 15 handshakes   vi 21 handshakes

   b The pattern is a summation of the natural numbers 1, 1 + 2, 1 + 2 + 3, 1 + 2 + 3 + 4 … This series of numbers is known as the triangular numbers.

Practice questions 2.2.7

1 a 4   b 60   c 9

2 Peter is 21.

3 24

4 24 marbles

5 €25.60

6 96 sweets

7 65 blueberries

Practice questions 2.2.8

1 55 numbers

2 271 numbers

3 140 metres

4 97 diagonals

5 The remainder is 1.

6 The food and water will last $7\frac{1}{2}$ days.

   Explanation:

   20 people have enough food and water for 12 days.

   3 days later, 20 people have enough food and water for 9 days, so one person would have enough for (20 × 9 =) 180 days.

   There are now 24 people, so there is enough food and water for (180 ÷ 24 =) $7\frac{1}{2}$ days.

Check your knowledge questions

1 Kelly is 11 and Calvin is 8.

2 Kelly earns more with Option 2 (€51.15) than she does with Option 1 (€20).

3 10 handshakes

4 1 ticket, 1 popcorn, 1 drink and 2 chocolate bars

5 72 seats

6 96 ways

7 15:35

8 They will meet at 18:08.

Chapter 3 answers

Do you recall?

1 Real numbers are all rational and irrational numbers, which includes positive and negative numbers, decimal numbers and fractions.

2 Student's own answer, for example, partitioning within the grid method

3 Add and subtract the numerator.

Practice questions 3.1

1 a 5472 b 0   c 571 d 56

    e 572   f 127   g 28   h 28

    i 0   j 5894

2 a True, 99 = 99

   b True, 180 = 180

   c Not true, 5 ≠ −5

   d True, 438 = 438

   e True, 712 = 712

   f Not true, $5 \neq \frac{1}{5}$

   g Not true, 55 ≠ −55

   h True, 647 = 647

   i Not true, 647 ≠ −647

   j True, 0 = 0

   k Not true, 25 ≠ 0.04

   l Not true, 237 ≠ 236

3 a 6871 b 472   c 0   d 0

    e 49   f 17   g 74   h 1

    i 6   j 97   k 0   l 14

4 a True b True c True d False

    e True f False g True h True

    i False j True

5 a 100   b 30   c 1000 d 50

    e 90   f 50

6 a 150   b 0   c 248 d 3457

    e 3700 f 1900 g 597 h 200

    i 0   j 1300

7 a 1200 b 1200 c 1200 d 1200
   e 1200 f 1200

The answers are the same because we could write every part of this question as
$3 \times 4 \times 10 \times 10$

8 a True   b True   c False
   d True   e True   f True

9 a 123×0 b 567×1

10 $3 \times 7 = 21 = 7 \times 3$ Hence, multiplying by 3 and 7 is the same as multiplying by 7 and 3 and is the same as multiplying by 21.

11 a $8 \times 3 = 24$, $24 \times 7 = 168$
    b $9 \times 3 = 27$, $27 \times 7 = 189$
    c $7 \times 9 = 63$, $63 \times 2 = 126$ or $7 \times 6 = 42$, $42 \times 3 = 126$
    d $13 \times 3 = 39$, $39 \times 5 = 195$
    e $9 \times 7 = 63$, $63 \times 5 = 315$
    f $11 \times 4 = 44$, $44 \times 7 = 308$

12 $3 \times 5 = 15$, so dividing by 3 then dividing by 5 is the same as dividing by 15

13 a $480 \div 3 = 160$, $160 \div 5 = 32$
    b $525 \div 3 = 175$, $175 \div 5 = 35$
    c $540 \div 3 = 180$, $180 \div 4 = 45$
    d $475 \div 5 = 95$, $95 \div 5 = 19$
    e $315 \div 3 = 105$, $105 \div 7 = 15$
    f $630 \div 7 = 90$, $90 \div 5 = 18$

14

6	8	2	96
4	9	7	252
1	3	5	15
24	216	70	

Practice questions 3.2

1 a 4, 8, 12, 16, 20, 24
   b 11, 22, 33, 44, 55, 66
   c 2, 4, 6, 8, 10, 12
   d 10, 20, 30, 40, 50, 60
   e 5, 10, 15, 20, 25, 30

2 a 1, 2, 3, 6
   b 1, 2, 3, 4, 6, 12
   c 1, 3, 5, 15
   d 1, 2, 4, 8, 16
   e 1, 5, 25
   f 1, 2, 4, 5, 10, 20, 25, 50, 100
   g 1, 2, 3, 5, 6, 10, 15, 30
   h 1, 2, 4, 5, 8, 10, 20, 40
   i 1, 3, 9, 27
   j 1, 3, 9, 11, 33, 99

3 a True b True c False d True
   e True f False g False h True

4 a 1, 2, 3, 4, 6, 8, 12, 24
   b 1, 2, 3, 4, 6, 9, 12, 18, 36
   c 1, 2, 3, 4, 6, 12
   d 12

5 a 8   b 5   c 10   d 7
   e 8   f 50   g 24   h 1

6 a 6, 12, 18, 24, 30, 36, 42, 48, 54, 60
   b 10, 20, 30, 40, 50, 60, 70, 80, 90, 100
   c 30, 60 (90, …)
   d 30

7 a 24   b 36   c 75   d 40
   e 72   f 44   g 42   h 60

8 a 52, 56, 60, 64, 68
   b 12, 15, 18, 21, 24, 27
   c 1, 2, 3, 5, 6, 10, 15, 30
   d 15, 30, 45, 60, 75, 90
   e 42, 49, 56, 63, 70, 77
   f 21, 42, 63

9 a 4, 10   b 10, 25, 40, 50, 60
   c 4, 10, 25, 50   d 16, 32, 40

10 a 10 is a factor of 30 and 40, not a multiple.
    b 120

11 2 and 5

12 $1 + 2 + 3 + 5 + 6 + 9 = 26$

13 10 and 16

14 a One of 1, 2, 3 or 6

   b One of 5, 10 or 40

   c Any multiple of 24, e.g. 24, 48, 72, …

15 There are many possible answers, e.g. 15 and 30, 15 and 45, 30 and 45, …

16 One of 3 and 5, 3 and 15 or 5 and 15

17 3 and 6

18 18 packs in total; each pack contains 2 pens and 3 pencils

19 360 seconds (6 minutes)

20 a

	Multiples of 2	Factors of 18	Multiples of 5
Even numbers	2, 4, 6, 8, 10, 12, …	2, 6, 18	10, 20, 30, …
Odd numbers		1, 3, 9	5, 15, 25, 35, …
Multiples of 3	6, 12, 18, 24, …	3, 9, 18	15, 30, 45, 60, …

   b Multiples of 2 cannot be odd numbers as they are all even numbers.

21 a A number is divisible by 8 if the last three digits are divisible by 8.

   b A number is divisible by 11 if the sum of the digits in odd-numbered places is equal to the sum of the digits in even-numbered places, or has a difference that is a multiple of 11.

   c A number is divisible by 25 if the last two digits are 00, 25, 50 or 75.

Practice questions 3.3.1

1 a 4^2 b 7^3 c 11^5 d 25^3

   e 2^6 f 8^4 g 9^5 h 12^4

2 a $4 \times 4 \times 4$

   b $5 \times 5 \times 5 \times 5$

   c $10 \times 10 \times 10 \times 10 \times 10 \times 10$

   d $7 \times 7 \times 7 \times 7 \times 7$

   e $3 \times 3 \times 3 \times 3 \times 3 \times 3$

   f $6 \times 6 \times 6 \times 6 \times 6 \times 6 \times 6$

   g $15 \times 15 \times 15 \times 15$

   h $9 \times 9 \times 9$

3 a 25   b 169   c 256

   d 1764   e 121   f 625

4 a 8   b 125   c 1000

   d 3375   e 5832   f 13 824

5 a 128   b 243   c 1296

   d 3125   e 248 832   f 20 736

6 3^4 is greater as $3^4 = 81$ and $4^3 = 64$

7 a 20   b 4000   c 700 000

   d 30 000   e 40 000   f 9 000 000

8 a 25   b 200   c 8

   d 9   e 64   f 1

   g 14   h 52   i 21

9 a 4   b 6   c 9   d 2

   e 5   f 3   g 1   h 8

10 a 8   b 6

11 a $1 + 9 \ (1^2 + 3^2)$ b $4 + 16 \ (2^2 + 4^2)$

   c $9 + 16 \ (3^2 + 4^2)$

12 8 and 6

13 4 and 8 as $4^3 = 64$ and $8^2 = 64$

14 $\dfrac{9^2 - 3^2}{1 + 2} = 24$

15 $7^2 - 3^2 = 40$ but $(7 - 3)^2 = 4^2 = 16$

16 a 125   b $66 \ (6^3 - 150)$

17 3, 1, 15, 34, 2

Practice questions 3.3.2

1 a $2^3 \times 2^4 = 2^7$ b $6^5 \times 6^3 = 6^8$

   c $7^2 \times 7^9 = 7^{11}$ d $9 \times 9^4 = 9^5$

   e $3 \times 3^1 = 3^2$ f $4^3 \times 4^5 = 4^8$

2 a 4^5 b 3^5 c 5^{11} d 2^{12}

   e 8^{13} f 3^{22} g 10^{21} h 12^{13}

3 She has multiplied the indices 3 and 4 and not added; $5^3 \times 5^4 = 5^7$

4 a $10^4 \times 10^{-1} = 10^3$ b $10^2 \times 10^{-3} = 10^{-1}$

 c $10^5 \div 10^2 = 10^3$ d $10^6 \div 10^4 = 10^2$

 e $10^5 \div 10^{-3} = 10^8$ f $\dfrac{10^7}{10^3} = 10^4$

 g $\dfrac{10^2}{10^6} = 10^{-4}$ h $\dfrac{10^2}{10^{-3}} = 10^5$

5 a 4^3 b 4^5 c 5^2 d 6^2

 e 2^3 f 4^{13} g 4^{12} h 6^4

 i 8^6 j 5^{10} k 7^5 l 2^3

6 a Student's own answer, for example, $6^7 \times 6^5$

 b Student's own answer, for example, $4^{10} \times 4^5$

 c Student's own answer, for example, $12^{10} \div 12^7$

 d Student's own answer, for example, $5^{12} \div 5^4$

 e Student's own answer, for example, $4^3 \div 4^6$

 f Student's own answer, for example, $3^2 \div 3^{10}$

 g Student's own answer, for example, $4^{10} \times 4^{-14}$

7

×	3^2	3^3	3^5	3^7	3^8
3^{-3}	3^{-1}	1 or 3^0	3^2	3^4	3^5
3^{-4}	3^{-2}	3^{-1}	3^1	3^3	3^4
3^6	3^8	3^9	3^{11}	3^{13}	3^{14}
3^7	3^9	3^{10}	3^{12}	3^{14}	3^{15}
3^{-1}	3^1	3^2	3^4	3^6	3^7

8 a $2^3 \times 2^7 = 2^9$ b $6^5 \times 6^9 = 6^{14}$

 c $7^2 \times 7^9 = 7^{11}$ d $9 \times 9^2 = 9^3$

 e $3^8 \times 3^{-1} = 3^7$ f $4^9 \times 4^{-3} = 4^6$

 g $2^{10} \div 2^3 = 2^7$ h $5^6 \div 5^4 = 5^2$

 i $10^2 \div 10^4 = 10^{-2}$

Practice questions 3.4

1 a 9 b 1 c 9

 d 32, 33, 34, 35, 36, 38, 39

2 a 3, 2, 19, 29, 31, 13

 b 15, 21, 25, 27, 33, 9

3 a Any of: 1, 2, 4, 5 or 10

 b Any of: 8, 16, 24, 32, 40, ….

 c Either 23 or 29

4 Any number that is a multiple of 13 will have a common factor of 13 as well as 1.

5 a b c d
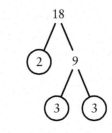
$18 = 2 \times 3 \times 3$ $12 = 2 \times 2 \times 3$

$25 = 5 \times 5$ Student's own tree diagram that gives $30 = 2 \times 3 \times 5$

6 a $2 \times 2 \times 5$ b $2 \times 5 \times 5$

 c $3 \times 3 \times 5$ d $3 \times 5 \times 5$

 e $2 \times 2 \times 2 \times 2 \times 3 \times 3$

 f $2 \times 2 \times 2 \times 3 \times 3$

 g $2 \times 2 \times 2 \times 2 \times 2 \times 2$

 h $2 \times 3 \times 3 \times 5$

7 a $2 \times 2 \times 3 \times 3 \times 5$

 b $3 \times 3 \times 5 \times 5$

 c $2 \times 2 \times 3 \times 3 \times 7$

 d $2 \times 2 \times 281$

 e $2 \times 2 \times 2 \times 2 \times 2 \times 2 \times 2 \times 2 \times 2$

 f $2 \times 2 \times 2 \times 2 \times 3 \times 3 \times 7$

 g $2 \times 2 \times 3 \times 3 \times 3 \times 5 \times 5$

 h $3 \times 3 \times 3 \times 3 \times 7 \times 7$

8 Either 11 and 19 or 13 and 17

9 9 is not a prime number. He has not finished. $9 = 3 \times 3$ so $27 = 3 \times 3 \times 3$

10 90

11 28

Practice questions 3.5.1

1. a. $\dfrac{1}{2} = \dfrac{\boxed{8}}{4} = \dfrac{5}{\boxed{10}} = \dfrac{\boxed{5}}{10} = \dfrac{8}{\boxed{16}}$

   b. $\dfrac{1}{10} = \dfrac{\boxed{4}}{40} = \dfrac{10}{\boxed{100}}$

   c. $\dfrac{1}{5} = \dfrac{\boxed{8}}{40} = \dfrac{5}{\boxed{25}}$

   d. $\dfrac{2}{3} = \dfrac{\boxed{4}}{6} = \dfrac{\boxed{10}}{15}$

   e. $\dfrac{3}{4} = \dfrac{\boxed{6}}{8} = \dfrac{9}{\boxed{12}} = \dfrac{24}{\boxed{32}}$

   f. $\dfrac{3}{2} = \dfrac{6}{\boxed{4}} = \dfrac{\boxed{18}}{12}$

2. a. $\dfrac{8}{14}$ b. $\dfrac{5}{18}$

   c. $\dfrac{5}{20}$ d. $\dfrac{1}{4}$

   e. $\dfrac{3}{4}$ f. $\dfrac{20}{200}$

3. $\dfrac{1}{6} = \dfrac{3}{18} = \dfrac{5}{30} = \dfrac{7}{42} = \dfrac{10}{60}$, $\dfrac{2}{3} = \dfrac{4}{6} = \dfrac{20}{30} = \dfrac{22}{33}$,

   $\dfrac{4}{5} = \dfrac{16}{20} = \dfrac{20}{25} = \dfrac{44}{55} = \dfrac{80}{100}$ and $\dfrac{1}{10} = \dfrac{5}{50} = \dfrac{10}{100}$

   $= \dfrac{20}{200} = \dfrac{100}{1000}$

   The odd one out is $\dfrac{50}{100}$

4. a. $\dfrac{4}{5}$ b. $\dfrac{2}{5}$

   c. $\dfrac{1}{2}$ d. $\dfrac{1}{4}$

   e. $\dfrac{9}{20}$ f. $\dfrac{4}{7}$

   g. $\dfrac{1}{4}$ h. $\dfrac{7}{10}$

5. $\dfrac{13}{45}, \dfrac{33}{35}, \dfrac{89}{100}$ and $\dfrac{15}{22}$ These cannot be simplified because the highest common factor of the numerator and denominator of each fraction is 1.

6. Hamid and Aref are both correct. However, Aref's method is more efficient.

7. a. < b. > c. = d. <

   e. = f. > g. < h. >

8. a. $\dfrac{1}{4}, \dfrac{4}{10}, \dfrac{1}{2}$ b. $\dfrac{3}{10}, \dfrac{1}{3}, \dfrac{2}{5}$

   c. $\dfrac{2}{10}, \dfrac{5}{20}, \dfrac{30}{100}$ d. $\dfrac{11}{20}, \dfrac{6}{10}, \dfrac{2}{3}, \dfrac{67}{100}$

   e. $\dfrac{1}{2}, \dfrac{13}{20}, \dfrac{3}{4}, \dfrac{8}{10}$

9. $\dfrac{\boxed{1}}{4}, \dfrac{1}{3}, \dfrac{\boxed{1}}{2}, \dfrac{3}{\boxed{5}}, \dfrac{3}{\boxed{4}}, \dfrac{8}{10}$

10. a. $\dfrac{2}{3}$ or $\dfrac{3}{4}$ or any fraction with a denominator of 1 or any fraction with a numerator of 4

    b. $\dfrac{1}{4}$ or $\dfrac{1}{3}$

11. a. $2\dfrac{1}{2}$ b. $1\dfrac{9}{10}$ c. $2\dfrac{4}{5}$

    d. $5\dfrac{5}{8}$ e. $7\dfrac{6}{7}$ f. $3\dfrac{49}{100}$

12. a. $\dfrac{7}{3}$ b. $\dfrac{19}{10}$ c. $\dfrac{12}{5}$

    d. $\dfrac{47}{11}$ e. $\dfrac{76}{7}$ f. $\dfrac{151}{7}$

13. a. 15 b. 10 c. 6 d. 5

    e. 20 f. 18 g. 25 h. 21

14. a. 5 b. 20 c. 4 d. 60

    e. 18 f. 14 g. 900 h. 45

15. a. \$2.50 b. \$0.40

    c. 1.8 kg d. 2.2 m

    e. \$0.90 f. 42 minutes

16. a. 4680 ml

    b. 4800 people

    c. 450 minutes or 7.5 hours

    d. 18 litres

    e. 21 600

17. 36

18. a. > b. < c. > d. =

    e. < f. >

19. 42

20. 75

21. 135

22. $\dfrac{1}{10}$ of her annual allowance = 12 lots of $\dfrac{1}{10}$ her monthly allowance. This is $\dfrac{12}{10}$ or $\dfrac{6}{5}$ of one month. As $\dfrac{6}{5} > \dfrac{2}{3}$, Zahra's better option is to take the $\dfrac{1}{10}$ of her annual allowance reward.

Practice questions 3.5.2

1 a $\frac{3}{10}$ b $\frac{4}{5}$ c $\frac{13}{15}$ d $1\frac{1}{8}$

 e $\frac{5}{6}$ f $\frac{11}{12}$ g $\frac{17}{20}$ h $1\frac{7}{30}$

2 a $\frac{1}{2}$ b $\frac{3}{10}$ c $\frac{1}{2}$ d $\frac{1}{10}$

 e $\frac{5}{24}$ f $\frac{1}{12}$ g $\frac{7}{30}$ h $\frac{19}{40}$

3 a $\frac{9}{10}$ b $\frac{49}{50}$ c $\frac{13}{50}$ d $1\frac{7}{100}$

 e $\frac{7}{24}$ f $\frac{5}{24}$ g $2\frac{1}{28}$ h $\frac{57}{100}$

4 a $2\frac{1}{2}$ b $3\frac{13}{40}$ c $4\frac{1}{2}$

 d $3\frac{23}{24}$ e $9\frac{13}{14}$ f $12\frac{39}{100}$

 g $20\frac{3}{20}$ h $11\frac{13}{30}$

5 a $\frac{5}{6}$ b $4\frac{5}{6}$ c $\frac{7}{10}$ d $2\frac{11}{12}$

 e $\frac{19}{30}$ f $6\frac{12}{25}$ g $1\frac{19}{24}$ h $1\frac{151}{200}$

6 a $\frac{9}{50}$ b $\frac{10}{21}$ c $\frac{12}{35}$ d $\frac{3}{10}$

 e $\frac{3}{7}$ f $\frac{1}{6}$ g $\frac{4}{15}$ h $\frac{1}{4}$

7 a $1\frac{1}{5}$ b $2\frac{1}{2}$ c 1 d $\frac{15}{16}$

 e $1\frac{5}{11}$ f 25 g $2\frac{1}{2}$ h $1\frac{1}{15}$

8 a 1 b 1 c $1\frac{13}{20}$ d $1\frac{7}{8}$

 e $1\frac{1}{5}$ f $2\frac{17}{30}$ g $18\frac{1}{3}$ h $\frac{5}{12}$

9 a 20 b $8\frac{3}{4}$ c $7\frac{1}{5}$ d $5\frac{5}{7}$

 e $\frac{13}{25}$ f $\frac{3}{4}$ g $2\frac{1}{2}$ h $1\frac{2}{7}$

10 $\frac{19}{30}$

11 a $\frac{17}{24}$ b $\frac{7}{24}$

12 $\frac{7}{12} + \frac{\boxed{4}}{7} = 1\frac{13}{84}$

13 $\frac{4}{21}$ m²

14 $\frac{9}{20}$

15 $\frac{11}{40}$ km

16 a $11\frac{7}{12}$ m b $6\frac{7}{8}$ m²

17

18 There are multiple answers, such as:

 $6\frac{3}{10} + 8\frac{7}{25} + 5\frac{21}{50}$

19 $-1\frac{7}{8}$

×	$1\frac{1}{3}$	$1\frac{1}{5}$	$2\frac{1}{4}$
$1\frac{2}{3}$	$2\frac{2}{9}$	2	$3\frac{3}{4}$
$1\frac{3}{5}$	$2\frac{2}{15}$	$1\frac{23}{25}$	$3\frac{3}{5}$
$2\frac{3}{4}$	$3\frac{2}{3}$	$3\frac{3}{10}$	$6\frac{3}{16}$

21 a $\frac{2}{5} + \frac{3}{10} = \frac{7}{10}$ b $1\frac{1}{8} + 1\frac{1}{3} = 2\frac{11}{24}$

 c $1\frac{1}{3} - \frac{3}{10} = 1\frac{1}{30}$ d $1\frac{1}{8} \times 1\frac{1}{3} = 1\frac{1}{2}$

 e $\frac{2}{5} \times \frac{3}{10} = \frac{3}{25}$ f $\frac{2}{5} - \frac{3}{10} = \frac{1}{10}$

Practice questions 3.6.1

1 a 9 b 4 c 25 d 17

 e 46 f 1 g 3 h 245

2 a 5.4 b 8.7 c 36.6 d 2.5

 e 0.1 f 0.1 g 10.0 h 458.1

3 a 254.3 b 0.59 c 2.2 d 15.15

 e 10.0 f 8.8 g 0.51 h 3.53

4 a 23.25 b 4587.74

 c 129.50 d $23\,564\,821.30$

 e 90.55 f 2300.72

5 a 23 b 4588

 c 129 d $23\,564\,821$

 e 91 f 2301

6 a 128 cm   b 2846 ml

   c 475 mm   d 5084 m

   e 56 113 g

7 13.99 and 13.95 both round to 14.0
13.94, 13.9 and 13.908 all round to 13.9

8 Any number between 14.85 and 15.4$\dot{9}$

9 a Any number from 7.75 to 7.79

   b Any number from 7.80 to 7.84$\dot{9}$

10 Four numbers between 3.445 to 3.454$\dot{9}$

11 23.25 ⑨ ⑨ [any digit]

12 a 3333\frac{1}{3}$

   b It is not possible to get $$\frac{1}{3}$ accurately. To
the nearest cent, this is £0.33 and 3 × $0.33
= $0.99
To split the $10 000, one would have to
receive $3333.34 and the other two receive
$3333.33.

13 a 19.54$\dot{9}$ b 18.45

Practice questions 3.6.2

1 a 5.61   b 4.085   c 258.95

   d 0.2   e 0.84   f 6.9

   g 97.109   h 12   i 20

   j 0.5   k 0.49   l 7000

2 a 0.54   b 0.11

   c 13.72   d 94

   e 0.0121   f 16

   g 12.3   h 0.0025

3 No, 0.58 is 10 times smaller than 5.8.
However, 270 is 100 times larger than 2.7, so
the answer will be 10 times larger overall.

4 a True   b False   c True

   d True   e True   f False

5 a 7 × 0.305 = 2.135

   b 0.07 × 305 = 21.35

   c 0.7 × 3050 = 2135

   d 0.07 × 30.5 = 2.135

   e 70 × 0.305 = 21.35

   f 0.0007 × 305 = 0.2135

6 Student's own answers, for example,
56 × 0.18 = 10.08, 56 × 18 = 1008,
0.56 × 0.018 = 0.01008

7 0.38 minutes

8 No, she has only 6.4 m

9 $15.25

10 a 14.6 cm   b 12 cm²

11 71.1 cm²

12 230.9 cm

13
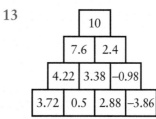

14

×	0.2	0.3	3	1.2
0.7	0.14	0.21	2.1	0.84
5.7	1.14	1.71	17.1	6.84
5.8	1.16	1.74	17.4	6.96

Check your knowledge questions

1 a 32 876   b 0   c 805

2 a True   b False   c True   d False

3 a 1   b 0   c 59   d 27

4 a 280   b 73

5 a 350   b 0   c 384

6 a 3, 6, 9, 12, 15, 18

   b 6, 12, 18, 24, 30, 36

   c 15, 30, 45, 60, 75, 90

7 a 1, 2, 5, 10

   b 1, 2, 3, 6, 9, 18

   c 1, 7, 49

   d 1, 2, 4, 5, 8, 10, 16, 20, 40, 80

8 a False   b True   c False   d True

9 52

10 90

11 a   2, 4, 8, 16   b   6, 12, 15, 36

    c   4, 16, 36   d   2, 11

12 a 4^3 b 9^5

13 a $7 \times 7 \times 7 \times 7$

    b $2 \times 2 \times 2 \times 2 \times 2 \times 2 \times 2$

14 a   36   b   144   c   169

15 a   64   b   343   c   1728

16 a   729   b   128   c   50 625

17 a   600   b   800 000

    c   250 000   d   49 000

18 a   34   b   432   c   125   d   125

19 a   5   b   12   c   4   d   10

20 36 and 64

21 a 7^7 b 11^{15} c 10^2

    d 10^{-2} e 3^3 f 10^{16}

    g 5^7 h 13^5 i 6^3

22 a $2 \times 2 \times 3 \times 3$ b $2 \times 2 \times 5 \times 5$

    c $2 \times 3 \times 5 \times 5$

23 $\dfrac{24}{35}$

24 a $\dfrac{3}{4}$ b $\dfrac{5}{6}$ c $\dfrac{17}{20}$ d $\dfrac{4}{5}$

25 a $>$ b $>$

26 $\dfrac{3}{5}, \dfrac{5}{8}, \dfrac{13}{20}, \dfrac{7}{10}$

27 $3\dfrac{1}{6}$

28 $\dfrac{15}{4}$

29 a   24   b   49   c   300

30 a   \$0.20   b   2.7 litres

    c   48 minutes

31 45

32 a $1\dfrac{26}{45}$ b $\dfrac{5}{24}$ c $\dfrac{4}{5}$ d $\dfrac{5}{6}$

    e $6\dfrac{1}{10}$ f $3\dfrac{19}{24}$ g $5\dfrac{2}{3}$ h $2\dfrac{3}{5}$

33 $2\dfrac{3}{4} - 1\dfrac{19}{20} = \dfrac{4}{5}$

34 a   15   b   8.0   c   0.36 d   2.3

35 227 mm

36 Any four numbers between 2.245 and $2.254\dot{9}$

37 a   159.78   b   1.26

    c   2.459   d   0.7

38 a   0.29   b   7.72   c   24.9001

39 a $0.04 \times 703 = 28.12$

    b $0.4 \times 7030 = 2812$

    c $0.0004 \times 703 = 0.2812$

Chapter 4 answers

Do you recall?

1 D shows a quarter turn.

2 A half turn, B three-quarter turn, C full turn

3 B

4 B

5 90°

6 360°, 180°

7 North, South, East and West

Practice questions 4.1.1

1 a   Obtuse   b   Reflex   c   Acute

2 a   49°, 72°, 27°   b   94°, 127°

    c   217°, 249°, 294°

3 i   a   acute   b   50°

   ii a   obtuse   b   110°

   iii a   acute   b   65°

   iv a   obtuse   b   145°

   v a   obtuse   b   98°

   vi a   acute   b   37°

4 a   She has read the wrong scale on the protractor.

    b   120°

5 a   340°   b   270°

    c   235°   d   323°

6 230°

7   a $x = 40°$, $y = 110°$, $z = 30°$

    b $a = 130°$, $b = 92°$, $c = 70°$, $d = 68°$

Practice questions 4.1.2

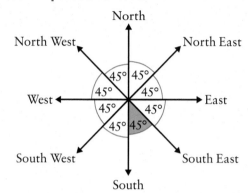

1

2 i   a   Acute   b

  ii   a   Obtuse   b

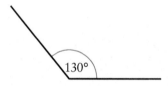

  iii a   Acute   b

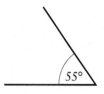

  iv a   Obtuse   b

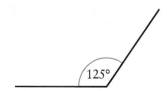

  v   a   Obtuse   b

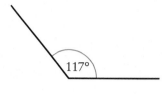

  vi a   Acute   b

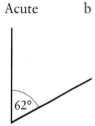

3   He has used the wrong scale on the protractor.

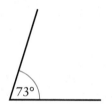

4

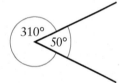

5   a

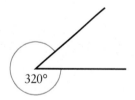

  b

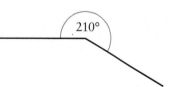

  c

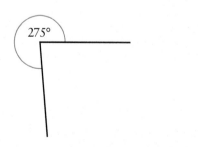

d

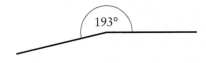

6

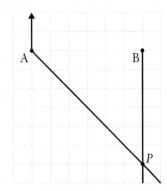

7   Student's own accurate drawings

8   a

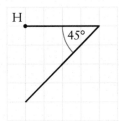

   b   4 km

   c   13.66 km (2 d.p.)

Practice questions 4.2.1

1   a   ∡PQR or ∡RQP   b   ∡CBD or ∡DBC

    c   ∡EFG or ∡GFE   d   ∡KLM or ∡MLK

    e   ∡UST or ∡TSU   f   ∡WZX or ∡XZW

2   a   ∡ABC or ∡CBA   b   ∡DGF or ∡FGD

    c   ∡KNL or ∡LNK   d   ∡PSR or ∡RSP

3   Max and Saleem are correct. The angle is at vertex D, and so D has to be the 'middle' letter in the angle name.

4   a   Angles can be drawn in any orientation.

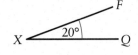

b

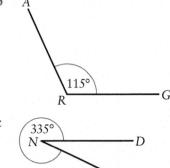

c

5   a   Right angle   b   Obtuse   c   Acute

6   a   85°   b   110°   c   60°

Practice questions 4.2.2

1   a $a = 65°$ b $b = 105°$ c $c = 20°$

    d $d = 31°$ e $e = 120°$ f $f = 170°$

    g $g = 45°$ h $x = 63°, y = 98°$

2   a   ∡$BCD = 150°$ (angles around a point add to 360°)

    b   i   ∡$EFG = 132°$ (supplementary angles, or angles on a straight line add to 180°)

        ii   ∡$FGI = 40°$ (supplementary angles, or angles on a straight line add to 180°)

3   a   i   No, two acute angles together are < 180°

        ii   Yes, two acute angles can add to 90°

    b   No, two obtuse angles together are > 90°

    c   No, a reflex angle itself > 180°

4   a $k = 45°$ (complementary angles)

    b $m = 36°$ (angles on a straight line add to 180°)

    c $n = 45°$ (angles on a straight line add to 180°)

    d $p = 94°$ (angles on a straight line add to 180°)

        $q = 66°$ (angles around a point add to 360°)

5 a 70° and 110°, 95° and 85°, 36° and 144°,
     165° and 15°, 105° and 75°

   b 20° and 160°, 154° and 26°

6 ∡FCA = 70° (angles on a straight line add to
   180°)

   ∡ACB = 35° (angles on a straight line add to
   180°)

   ∡FAC = 2∡ACB

7 Student's own answers

8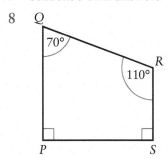

Practice questions 4.3.1

1 a Isosceles, obtuse angled

   b Right-angled, scalene

   c Right-angled, isosceles

   d Equilateral (acute angled)

   e Obtuse-angled, scalene

   f Equilateral

2 a Isosceles, obtuse angled

   b Scalene, right-angled

   c Equilateral

   d Scalene, acute angled

3 a AB and AC are equal.

   b ∡ACB and ∡ABC are equal.

   c Isosceles (acute angled)

   d Right-angled (scalene)

   e AD is a line of symmetry; ABD and ACD
     have the same angles and the same side
     lengths and are identical.

4 Student's own diagrams. Isosceles (possibly
   equilateral depending on points chosen).
   The two radii are always equal length, so the

triangle always has two equal sides.

Practice questions 4.3.2

1 a 10°   b 40°   c 75°
   d 96°   e 35°   f 22°

2 60°

3 a g = 63°, h = 54°

   b j = 70°, k = 70°

   c m = 25°, n = 130°

   d p = 45°, q = 45°

4 17°

5 a Student's own diagram   b 36°

6 Not all the reasons are given here, as students
   may find the angles in different steps.

   a x = 20°, angles in a triangle add up to 180°
     y = 160°, angles along a straight line add
     up to 180°

   b z = 84°, ∡EFG = 42°, isosceles triangle
     ∡EGF = 96° (180° − 84°), angles along a
     straight line add up to 180°

   c c = 130°, angles along a straight line add
     up to 180°, d = 50°, angles along a straight
     line add up to 180°, e = 80°, ∡RST = 50°,
     isosceles triangle, making ∡RTS 80°

7 48° and 66° or 57° and 57°

8 45° (angle is the vertex of an isosceles, right-
   angled triangle)

Practice questions 4.3.3

1 Student's accurate drawings of triangles

2 Student's accurate drawing. LN ≈ 2.2 cm,
   MN ≈ 4 cm

3 Student's accurate drawing of triangle

4 Student's accurate drawing of triangle

5 Student's accurate drawing of triangle. Length
   of third side = 5 cm

6 Student's accurate drawing. Tree ≈ 25.7 m

7 Student's own answers

8 Student's accurate drawing. $AB \approx 11.1\,\text{km}$
   1 cm on diagram represents 1 km.

9

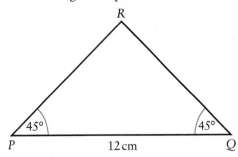

$RP = RQ \approx 8.5\,\text{km}$

10 Student's accurate drawing. 7.1 km (approx.)

Check your knowledge questions

1 a i Obtuse   ii Acute   iii Acute
   b i 150°   ii 85°   iii 34°

2 220°

3 a South   b North

4 Student's accurate drawings of angles

5 a

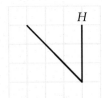

   b 3 km

6 a ∡PSR or ∡RSP
   b ∡XZY or ∡YZX
   c ∡ABC or ∡CBA
   d ∡GFH or ∡HFG

7 a i Acute   ii Obtuse
   b i 92°   ii 50°

8 a $a = 75°$
   b $b = 132°$
   c $c = 40°$
   d $d = 150°$
   e $e = 30°$
   f $f = 155°$, $g = 25°$, $h = 155°$

9 a 120°   b 30°

10 a Isosceles, acute angled
    b Equilateral
    c Right-angled, scalene
    d Isosceles, obtuse angled

11 180°

12 73°

13 55°

14 Student's accurate drawing. $TU \approx 3.6\,\text{cm}$

15 Student's accurate drawings of two isosceles
    triangles. One with angles 46°, 46°, 88°, and
    one with angles 46°, 67°, 67°

16

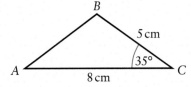

$\measuredangle ABC \approx 71°$

17

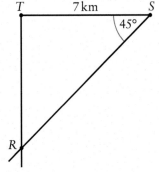

$RT = 7\,\text{km}$

Chapter 5 answers

Do you recall?

1 Student's own answer, for example, divide the
   numerator by the denominator. $\dfrac{3}{100} = 0.03$

2 Student's own answer, for example, divide the
   numerator and denominator by their highest
   common factor. $\dfrac{80}{100} = \dfrac{4}{5}$

3   Student's own answer, for example, add the decimal point after the whole number using the bus stop method. $4)\overline{3.{}^30{}^20}$ with 0.75 above

4   Student's own answer, for example, divide the numerator by the denominator to give the whole number and the remainder is the numerator; the denominator stays the same $\frac{7}{5} = 1\frac{2}{5}$

5   Student's own answer, for example, multiply the numerators together and the denominators together and then simplify if possible $\frac{3}{5} \times \frac{10}{21} = \frac{30}{105} = \frac{2}{7}$

Practice questions 5.1.1

1   a   Terminating   b   Recurring
   c   Recurring   d   Terminating
   e   Recurring   f   Terminating
   g   Recurring   h   Recurring
   i   Terminating   j   Terminating
   k   Terminating   l   Recurring

2   a $0.\dot{7}$ b $7.5\dot{2}$ c $0.\dot{2}\dot{5}$
   d $0.3\dot{6}$ e $9.\dot{8}\dot{0}$ f $3.1\dot{0}\dot{9}$
   g $0.\dot{1}4285\dot{7}$ h $14.562\dot{2}3\dot{4}$

3   a 0.7 b 0.9 c 0.03
   d 0.28 e 0.008 f 0.042
   g 0.125 h 0.67 i 0.34
   j 3.5 k 0.84 l 0.386

4   a 0.4 b 0.25 c 0.45 d 0.26
   e 0.125 f 0.625 g 0.875 h 0.44
   i $0.\dot{1}$ j $0.\dot{5}$ k $0.\dot{3}$ l $0.1\dot{6}$
   m $0.\dot{3}$ n $0.8\dot{3}$ o $0.\dot{6}$ p $0.\dot{7}$

5   a $2.\dot{1}$ b $7.0\dot{9}$ c $5.\dot{5}$
   d $12.3\dot{6}$ e 6.0625 f $2.\dot{7}1428\dot{5}$
   g $19.5\dot{3}$ h 1.425

6   0.01 and $\frac{1}{100}$, 0.3 and $\frac{3}{10}$, 0.8 and $\frac{4}{5}$, 0.5 and $\frac{1}{2}$, 0.24 and $\frac{12}{50}$, 0.75 and $\frac{3}{4}$, 0.4 and $\frac{2}{5}$, 0.35 and $\frac{7}{20}$

7   $\frac{3}{10}$, 0.33, $\frac{1}{3}$

8   She is incorrect, $\frac{3}{5}$ is the smallest as $\frac{3}{5} = 0.6$, $\frac{3}{4} = 0.75$ and $\frac{2}{3} = 0.\dot{6}$

9   Any decimal between $0.\dot{6}$ and 0.75

10 She is incorrect, multiplying by $\frac{1}{3}$ is the same as multiplying by $0.\dot{3}$ not 0.33

11 a   There are 60 minutes in 1 hour not 100 minutes, so 1 minute is $0.01\dot{6}$ hours not 0.01 hours.
   b   21 minutes

12 $\frac{2}{5} = 0.4$, $\frac{5}{20} = 0.25$, $\frac{16}{50} = 0.32$ and $\frac{1}{3} = 0.\dot{3}$
So, $\frac{16}{50}$ is the nearest to 0.3.

13 Any decimal between 0.85 and 0.92

14 a   Recurring decimals: $\frac{1}{3}, \frac{1}{6}, \frac{1}{7}, \frac{1}{9}, \frac{1}{11}, \frac{1}{12}, \frac{1}{13}$
     Terminating: $\frac{1}{2}, \frac{1}{4}, \frac{1}{5}, \frac{1}{8}, \frac{1}{10}$
   b   If the denominator has only prime factors of 2 and/or 5, then the fraction will produce a terminating decimal. If the denominator has any other prime factors, then the fraction will produce a recurring decimal.

Practice questions 5.1.2

1   a $\frac{70}{100}$ or $\frac{7}{10}$ b $\frac{77}{100}$ c $\frac{7}{100}$
   d $\frac{10}{100}$ or $\frac{1}{10}$ e $\frac{1}{100}$ f $\frac{29}{100}$
   g $\frac{41}{100}$ h $\frac{22}{100}$ i $\frac{33}{100}$

2   a $\frac{1}{4}$ b $\frac{1}{5}$ c $\frac{1}{2}$ d $\frac{9}{20}$
   e $\frac{13}{20}$ f $\frac{9}{10}$ g $\frac{3}{5}$ h $\frac{1}{20}$
   i $\frac{1}{50}$ j $\frac{2}{25}$

3 a 0.21 b 0.27 c 0.9 d 0.99

e 0.09 f 0.22 g 0.02 h 0.01

i 0.03 j 1.3 k 1.29 l 2.06

4 0.3%, 0.03, 3.3%, 30%, 0.33

5 a $1\frac{1}{2}$ b $3\frac{1}{4}$ c $1\frac{2}{5}$

d 5 e $4\frac{2}{5}$ f $2\frac{11}{50}$

g $1\frac{1}{4}$ h $2\frac{7}{100}$

6 a 7% b 12% c 41%

d 94% e 1% f 37%

g 2% h 15% i 52%

j 70% k 500% l 130%

7 $\frac{3}{4} = 75\%, \frac{3}{5} = 60\%, \frac{7}{10} = 70\%, 1\frac{1}{2} = 150\%,$

$\frac{39}{50} = 78\%, \frac{13}{100} = 13\%, \frac{3}{20} = 15\%$

8 a 80% b 55% c 25%

d 37% e 58% f $66.\dot{6}\%$

g $16.\dot{6}\%$ h 175%

9 a 47% b 85% c 30%

d 3% e 33% f 12.5%

g 170% h 325% i 437.5%

j 400% k 801% l 110%

10

Fraction	Decimal	Percentage
$\frac{7}{100}$	0.07	7%
$\frac{7}{20}$	0.35	35%
$\frac{7}{10}$	0.7	70%
$\frac{3}{100}$	0.03	3%
$1\frac{1}{4}$	1.25	125%
$\frac{2}{5}$	0.4	40%
$\frac{1}{6}$	$0.1\dot{6}$	$16.\dot{6}\%$
$\frac{1}{3}$	$0.\dot{3}$	$33.\dot{3}\%$
$\frac{9}{100}$	0.09	9%

$\frac{1}{8}$	0.125	12.5%
$1\frac{9}{10}$	1.9	190%

11 a $\frac{1}{40}$ and 0.025 b $\frac{13}{400}$ and 0.0325

c $\frac{19}{400}$ and 0.0475 d $\frac{81}{1000}$ and 0.081

e $\frac{27}{125}$ and 0.216 f $\frac{137}{1000}$ and 0.137

g $\frac{9}{400}$ and 0.0225 h $\frac{107}{10000}$ and 0.0107

12 a $\frac{49}{200}$ b 0.245

13 a $\frac{1}{30}$ b 0.1

14 65% is the greatest value as $\frac{3}{5} = 0.6 = 60\%$

15 a This is correct. $\frac{27}{50} = \frac{54}{100} = 54\%$

b This is not correct. $\frac{1}{20} = \frac{5}{100} = 5\%$ not 20%

c This is not correct. $\frac{3}{10} = \frac{30}{100} = 30\%$ not 3%

16 a $\frac{34}{50}$ or $\frac{17}{25}$ b 68%

17 85%

18 135.5%, 13.5, $13\frac{3}{5}$

Fraction	Decimal	Percentage
$\frac{1}{20}$	0.05	5%
$\frac{1}{10}$	0.1	10%
$\frac{1}{5}$	0.2	20%
$\frac{7}{50}$	0.14	14%
$\frac{3}{20}$	0.15	15%
$\frac{16}{25}$	0.64	64%
$\frac{2}{25}$	0.08	8%
$\frac{13}{20}$	0.65	65%

b 0.5, 1%, $\frac{1}{14}$

Practice questions 5.2

1. a 22 g   b 120 km   c 24 kg
   d 40 g   e 0.5 cm   f 0.9 m
   g 6 litres   h 28 kg   i 112 km
   j 10.5 hours   k 81 minutes
   l 27 seconds

2. a $112.50   b $1.50   c $18.00
   d $8.00   e $4.75   f $2.09
   g $7.02   h $0.44

3. 6 trading cards

4. 1860 worms

5. 3 hats

6. 45 m²

7. a $12 750   b $750

8. He should take 65% of $40 as this is $26. 15% of $140 is only $21

9. No, she does not qualify. 95% of 700 is 665 wins, so she is one win short.

10. The decimal equivalent to 8% is 0.08 not 0.8.

11. $3

12. 55% of 12 000 = 6600 and 70% of 9400 is 6580, so town A casts more votes

13. No, 20% of 25% is only 5% as 0.2 × 0.25 = 0.05

14. $72.60

15. Aisha has 350 stickers, Baris has 250 stickers and Camila has 300 stickers. There is a difference between the largest and smallest number of stickers of 100.

16. There are 13 points separating the highest and lowest scores. Bisma = 68 points, Cole = 55 points

Practice questions 5.3

1. a 25%   b 40%   c 84%
   d 20%   e 35%   f 65%

2. a 50% b 25% c 30% d 8%

e 25% f 35% g 5% h 60%

3. a 35%   b 10%   c 30%
   d 55%   e 90%   f 75%

4. 22%

5. 60%

6. 35%

7. 12.5%

8. 37.5%

9. a 46.4%   b 75%

10. 80%

11. 15%

12. No, she will not have to retake. $\frac{22}{30} = 73.3\%$

13. 78.4%

14. $\frac{26}{30}$

15. Yes, Paris does go through to the final round as $\frac{162}{225} = 72\%$

16. Mazin, $\frac{250}{2250} = 11.1\%$. Ola, $\frac{180}{1680} = 10.7\%$ Mazin has the higher percentage of fruit juice in her drink.

Practice questions 5.4

1. a Even-chance
   b Impossible
   c Student's own answer as this will depend on school policy
   d Certain
   e Student's own answer as this will depend on each individual student
   f Student's own answer as this will depend on each individual student
   g Certain
   h Unlikely/Impossible depending on student's definition of a superhero

10 Answers

2 Answers may vary from the above.
The diagrams should match the answers.
For example:

| Impossible | | Even-chance | | Certain |

b c a f d
h e g

3 a $\frac{1}{6}$ b $\frac{1}{6}$ c $\frac{1}{2}$

d $\frac{1}{2}$ e $\frac{1}{3}$ f 0

4 a $\frac{2}{3}$ b $\frac{1}{3}$ c $\frac{5}{6}$

d $\frac{1}{6}$ e 1 f 0

5 a $66.\dot{6}\%$ b $33.\dot{3}\%$ c $83.\dot{3}\%$

d $16.\dot{6}\%$ e 100% f 0%

6 Student's own diagram, with 4 white and 1 black counter or any multiple, for example, 8 white and 2 black

7 a i $\frac{3}{10}$ ii $\frac{1}{5}$ iii $\frac{1}{2}$

b The total of all three probabilities is 1.

8
0 $\frac{1}{2}$ 1
ii i iii

9 a $\frac{1}{2}$ b $\frac{1}{4}$ c $\frac{5}{12}$

d $\frac{1}{2}$ e $\frac{1}{12}$

10 The spinner should have one 1, two 2s, three 3s and two 4s.

11 $\frac{4}{7}$

12 $\frac{5}{16}$

13 a Bag A and P(R) = $\frac{1}{2}$. Bag B and P(R) = $\frac{1}{4}$

Bag C and P(R) = $\frac{5}{8}$. Bag D and P(R) = $\frac{7}{8}$

Bag E and P(R) = $\frac{1}{8}$.

b P(R) = $\frac{3}{4}$. This is Bag F.

c Raul should choose bag D. It has the highest probability of selecting a red counter and it is impossible to choose a green.

d Yousef should choose bag B. It has the highest probability of choosing a green counter.

14 a $\frac{3}{10}$ b $\frac{11}{15}$ c 0

15 a $\frac{8}{13}$ b $\frac{9}{13}$

c 100 There is one 20-euro note, four 10-euro notes and eight 5-euro notes making 100 euro.

16 Any multiple of 15 digits where $\frac{1}{5}$ are 1s, $\frac{3}{5}$ are 5s and $\frac{1}{15}$ each of 2, 3 and 4: e.g. 3 × 1, 1 × 2, 1 × 3, 1 × 4 and 9 × 5

17 P(E) = 0 as P(C) = 12 × 0.05 = 0.6, P(D) = 0.6 ÷ 3 = 0.2 and P(A) + P(B) +P(C) + P(D) = 1 so P(E) must be 0

18 Bag A and P(R or B) = $\frac{7}{8}$, Bag B and P(B or G) = $\frac{3}{4}$, Bag D and P(not G) = 1, Bag E and P(not R) = $\frac{7}{8}$, Bag F and P(B or G) = $\frac{1}{4}$

Note that Bag F also fits P(R or B) = $\frac{7}{8}$ but this would leave two bags left over. So Bag C is the one left over.

Check your knowledge questions

1 a Recurring b Terminating
c Recurring d Recurring

2 a $0.\dot{5}$ b $3.6\dot{4}$
c $0.1\dot{7}$ d $2.5\dot{1}6\dot{5}$

3 a 0.09 b 0.64 c 0.175
d 0.34 e 4.5 f 0.48
g 0.65 h 0.6 i 0.375
j $0.\dot{7}$ k $0.\dot{1}4285\dot{7}$ l $0.0\dot{9}$

4 a $2.\dot{3}$ b 2.1875
c $5.\dot{5}7142\dot{8}$ d $12.2\dot{6}$

5 0.6, 0.65, $\frac{2}{3}, \frac{7}{10}$

6 Any decimal between $0.\dot{3}$ and 0.4

7 $\frac{23}{50}$ is closest to 0.5; $\frac{3}{5}$ = 0.6 (0.1 too big), $\frac{9}{20}$ = 0.45 (0.05 too small), $\frac{23}{50}$ = 0.46 (0.04 too small), $\frac{4}{7}$ = $0.\dot{5}7142\dot{8}$ (0.0714285… too big)

8 a $\frac{2}{5}$ b $\frac{7}{20}$ c $\frac{19}{20}$ d $\frac{1}{25}$

9 a 0.37   b 0.3   c 0.96

   d 0.04   e 1.8   f 1.03

10 0.7%, 0.07, 7.7%, 70%, 0.77

11 a $2\frac{1}{2}$ b $1\frac{3}{4}$ c 3   d $3\frac{6}{25}$

12 a 3%   b 14%   c 65%   d 64%

   e 30%   f 170%   g 62%   h $83.\dot{3}\%$

13 a 23%   b 40%   c 1%

   d 37.5%   e 150%   f 207%

14

Fraction	Decimal	Percentage
$\frac{1}{25}$	0.04	4%
$\frac{17}{20}$	0.85	85%
$\frac{1}{10}$	0.1	10%
$\frac{3}{50}$	0.06	6%
$2\frac{3}{4}$	2.75	275%
$\frac{3}{5}$	0.6	60%
$\frac{5}{6}$	$0.8\dot{3}$	$83.\dot{3}\%$
$\frac{2}{3}$	$0.\dot{6}$	$66.\dot{6}\%$
$\frac{5}{8}$	0.625	62.5%
$1\frac{3}{10}$	1.3	130%

15 $\frac{7}{20}$ as $\frac{7}{20}$ = 0.35, which is greater than 33% = 0.33, which is greater than 0.3

16 a $\frac{3}{5}$ b 60%

17 224.5%, 24.4, $24\frac{4}{5}$

18 a 24 kg   b 75 ml

   c 2.8 m   d 2.5 hours

19 a $1.20   b $40.00

   c $1.95   d $3.30

20 21 students

21 Callum scored the highest marks. $\frac{7}{8}$ of 40 = 35, which is more than 85% of 40 = 34, which is more than 32 marks

22 a 75%   b 60%

23 60%

24 31.25%

25 a $\frac{1}{6}$ b $\frac{1}{2}$ c $\frac{1}{3}$ d 0

26 a $\frac{1}{4}$ b $\frac{1}{8}$ c $\frac{5}{8}$

27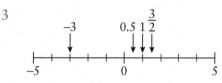

28 a $\frac{1}{2}$ b $\frac{5}{6}$ c 6

Chapter 6 answers

Do you recall?

1 a Multiplication and division first and addition and subtraction next

   b $5 + 4 \times 2 - 6 \div 3 = 5 + 8 - 2 = 11$

2 a Use brackets, because the inside of a bracket must be calculated first

   b $3 + 4 \times 2 = 3 + 8$, so 11

     $(3 + 4) \times 2 = 7 \times 2$ so 14

3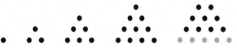

Practice questions 6.1

1 a

For instance, the base of each new triangle has one more dot than the previous one.

   b ⬤ ⬤ ⬤ ⬤ ⬤ ⬤

For instance, the cake is shared among a number of people that increases by one each time.

   c ⬜ ⬜⬜ ⬜⬜⬜ ⬜⬜⬜⬜ ⬜⬜⬜⬜⬜

For instance, each time, three sides of a square are added.

d

For instance, the base of each new triangle has one more dot than the base of the previous triangle.

2 a

Position in pattern	1	2	3	4	5
Number of dots	1	3	6	10	15

bi

Position in pattern	1	2	3	4	5
Number of slices	1	2	3	4	5

ii

Position in pattern	1	2	3	4	5
Size of slice	1	$\frac{1}{2}$	$\frac{1}{3}$	$\frac{1}{4}$	$\frac{1}{5}$

ci

Position in pattern	1	2	3	4	5
Number of squares	1	2	3	4	5

ii

Position in pattern	1	2	3	4	5
Number of segments	4	7	10	13	16

d

Position in pattern	1	2	3	4	5
Number of circles	1	4	9	16	25

3 a   i 3   ii 5   iii 7

b

Number of boxes	1	2	3	4	5	6	7
Number of visible faces	3	5	7	9	11	13	15

c   For instance, the number of visible faces starts from 3 and increases by two each time, or the number of visible faces equals the number of boxes plus 2.

Practice questions 6.2

1   a   5, 8, 11, 14

b   10, 8, 6, 4

c   1, 3, 9, 27

d   25, 5, 1, 0.2 (or $\frac{1}{5}$)

e   −10, −1, 8, 17

f   5, 2, −1, −4

g   2, −2, 2, −2

h   9, −3, 1, −$\frac{1}{3}$

i   1, 4, 13, 40

j   1, 6, 21, 66

k   3, 2, 0, −4

l   2, 4, 16, 256

2   a   Start with 7, add 7. 28, 35

b   Start with −1, subtract 3. −10, −14

c   Start with 4, multiply by 5. 500, 2500

d   Start with 2, subtract 2. −4, −6

e   Start with 10, divide by 2. 1.25, 0.625

f   Start with 4, multiply (or divide) by −1. −4, 4

g   Start with 5, multiply by 10. 2500, 12 500

h   Start with 2, add 2. 8, 10

i   Start with 2, multiply by 2. 16, 32

j   Start with 2, square each time. 256, 65 536

k   Start with −$\frac{1}{2}$, add 1. $\frac{5}{2}$, $\frac{7}{2}$

l   Start with 1, divide by 3 (or multiply by $\frac{1}{3}$). $\frac{1}{27}$, $\frac{1}{51}$

3   a   −5, 0

b   1, 0

c   0.2 (or $\frac{1}{5}$), 1

d   10 000, 1000

e   1, 1.5

f   1, $\frac{1}{2}$

g   29, 22

Practice questions 6.3

1   a 0   b 2   c 16   d 1

e −1   f −2   g 6   h 4

i 12   j 3   k 15   l 10

2   a −2   b 3   c −2   d 0

e 0   f 0   g 1   h 1

i −1   j 0   k 2

3   For instance:

a $A = s^2$

b $C = Nc$

c $P = a + b + c$ or $P = l_1 + l_2 + l_3$

d $S = A + B + C$ or $N_S = N_A + N_B + N_C$

e $s = \dfrac{d}{t}$

f $a = \dfrac{x + y}{2}$

g $F = L + 3$ or $a_F = a_L + 3$

4   a

s	1	2	3	4
w	3	4	5	6

b

r	0	2	4	6
q	0	4	8	12

c

x	1	2	3	4
y	9	8	7	6

d

t	1	2	3	4
l	0.5	1	1.5	2

e

s	1	2	3	4
w	3	5	7	9

f

x	0	1	2	3
y	10	8	6	4

g

x	1	2	3	4
y	6	6	4	0

h

x	1	2	3	4
y	24	12	8	6

i

x	0	1	2	3
y	0	1	4	9

5   a $y = x + 5$

b $b = a$

c $q + t = 3$ or $q = 3 - t$

d $d = 5w$

e $b = \dfrac{a}{2}$

f $y = 5x$

g $b = 5a - 1$

h $b = 3a$

i $xy = 12$ or $y = \dfrac{12}{x}$

j $b = -a$

6   a $c = 0.50h + 3.00$

b   €7.00

c   7 hours

7   a

s	3	6	9	12
n	12	14	16	18

b $n = 10 + 2 \times \dfrac{s}{3}$ or: starting from 10 white tiles when $s = 0$, then for every additional three green tiles there are two more white tiles.

c   22

d   15

8   a $n = \dfrac{2}{3}t$ or $t = 1.5n$

b   10th floor

c   22.5 seconds

Practice questions 6.4

1   a

n	1	3	5	7	9
Q	3	5	7	9	11

b

x	0	1	2	3	4	5	6	7
y	10.5	9	7.5	6	4.5	3	1.5	0

c

x	−6	−4	−2	0	2	4	6
y	9	8	7	6	5	4	3

d

n	0	2	4	6	8	10	12
p	0	1	2	3	4	5	6

2   a

b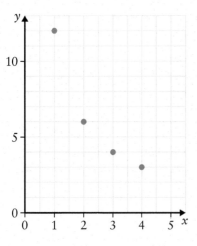

c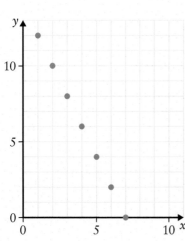

d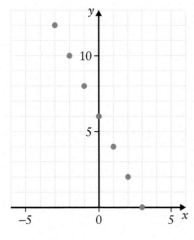

e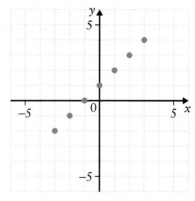

f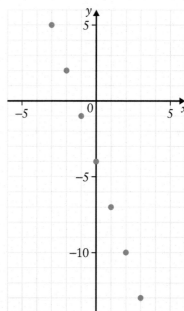

3 a

x	0	1	2	3	4	5
y	0	2	4	6	8	10

b

x	0	1	2	3	4	5
y	1	4	7	10	13	16

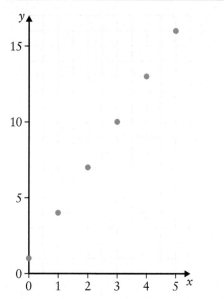

c

x	0	1	2	3	4	5
y	5	4	3	2	1	0

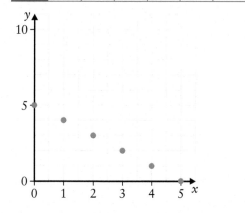

d

x	0	1	2	3	4	5
y	1	0.50	0.33	0.25	0.20	0.17

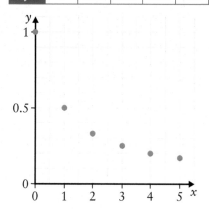

4 a $p = \dfrac{n}{2}$ b $Q = n + 2$

   c $y = 6 - \dfrac{x}{2}$ d $y = 10.5 - 1.5x$

5 a $c = 3 + 0.5h$

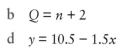

b $n = 10 + 2 \times \dfrac{s}{3}$

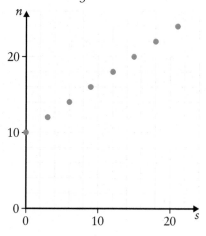

c $n = \frac{2}{3}t$

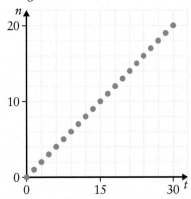

or $t = 1.5n$

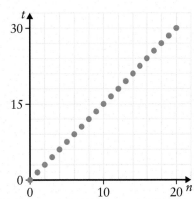

Check your knowledge questions

1   a

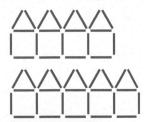

b

Position in pattern	1	2	3	4
Number of sticks	6	11	16	21

c   Start from 6, then add 5.
    or
    The number of sticks is 5 times the position in the pattern, plus 1.

d $s = 5n + 1$, where s is the number of sticks and n is the position in the pattern

2   a   2, 6, 10, 14

    b   5, −5, −15, −25

    c   2, 4, 8, 16

    d   3, −6, 12, −24

    e   2, 9, 30, 93

    f   2, 7, 22, 67

3   a   0, 3, 6, 9, 12

    b   3, 5, 7, 9, 11

    c   0.25 (or $\frac{1}{4}$), 1, 4, 16, 64

    d   0, 2, 4, 6, 8

    e   12, 8, 4, 0, −4

    f   −3, −5, −7, −9, −11

    g   4, −4, 4, −4, 4

    h   2, 4, 6, 8, 10

    i   2, 4, 8, 16, 32

    j   2, 4, 16, 256, 65 536

4   a   12   b   −1   c   0

    d   1   e   −6   f   3

    g   5   h   20   i   12

5   a

x	1	2	3	4
y	3	4	5	6

b

x	1	2	3	4
y	1	0	−1	−2

c

x	−3	−2	−1	0
y	−8	−6	−4	−2

d

x	1	2	3	4
y	1	0	−1	−2

e

x	0	0.5	1	1.5
y	2	1.5	1	0.5

f

x	−1.5	−1	−0.5	0
y	0.5	1	1.5	2

g

x	−1.5	−1	−0.5	0
y	2.5	2	1.5	1

6 a $y = x + 4$

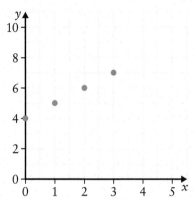

b $y = x + 3$

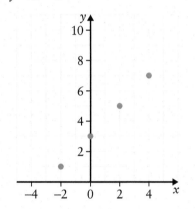

c $y = -9 - 2x$

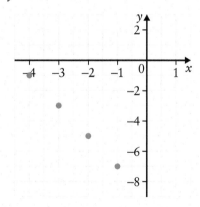

d $y = 1 - 3 \times \dfrac{x}{5}$

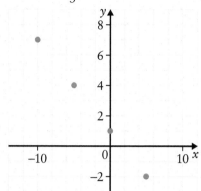

7 a

x	0	2	4	6	8	10
y	8	7	6	5	4	3

$y = 8 - \dfrac{x}{2}$

b

x	−6	−3	0	3	6
y	−3	−1	1	3	5

$y = 1 + \dfrac{2}{3}x$

c

x	−6	−3	0	3	6
y	7	6.5	6	5.5	5

$y = 6 - \dfrac{x}{6}$

d

x	−5	−3	−1	1	3	5
y	−4	−3	−2	−1	0	1

$y = -1.5 + \dfrac{x}{2}$

8 a

x	−4	−2	0	2	4
y	−7	−3	1	5	9

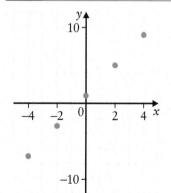

b

x	−4	−2	0	2	4
y	14	−8	−2	4	10

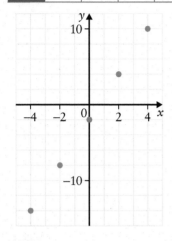

c

x	−4	−2	0	2	4
y	7	−5	−9	−5	7

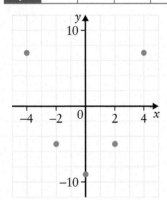

d

x	−4	−2	0	2	4
y	−1	−2	undef.	2	1

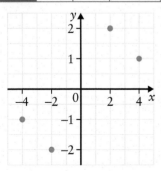

Chapter 7 answers

Do you recall?

1   A pattern is a regularity in a process that makes it possible to predict its outcomes. The rule for a pattern can be described in words, using an equation, or with a graph.

2   For instance, when a rule is expressed with an equation (an algebraic rule) we can solve for either variable and answer questions using few operations.

3   a   The first number in the pair – the abscissa – is placed on the x number line, and a vertical line is drawn through it. The second number in the pair – the ordinate – is placed on the y number line, and a horizontal line is drawn through it. The point is placed at the intersection of the two lines.

b

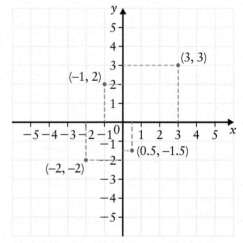

One point is placed on the Cartesian plane for each pair of numbers related to each other by the rule.

4 a Find some coordinates, e.g. when x is 0, y is 2; x is 3, y is 8, etc. Then plot the points on a graph.

   b

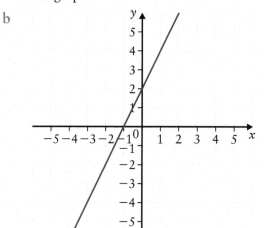

Practice questions 7.1

1 a $3g$ b $2n$ c $3y$

   d $14b$ e $-x$ f $-3g$

   g y h $2(y + z)$ i $\frac{1}{2}w, \frac{w}{2}$

   j $3a + 4b$ k $4a - 5b$ l $a + 10$

2 a $\frac{3}{8}$ b $\frac{n}{2}$ c $\frac{x}{3}$

   d $\frac{15}{b}$ e $\frac{2b}{3}$ f $\frac{x}{q}$

   g x or $\frac{x}{1}$ h $\frac{1}{x}$ i $\frac{2}{y + z}$

   j $\frac{x + 1}{x - 1}$ k $\frac{3}{2x}$

3 a 1024

   b 4

   c 2.42×10^{24}

   d 729

4 a $2 \cdot s + t$ b $2 \cdot (s + t)$

   c $3 \cdot (x - y)$ d $3 \cdot x - y$

   e $3 \cdot x \cdot y$ f $3 \cdot x \div y$

   g $2 \cdot x - 4 \cdot y$ h $2 \cdot (x - 4 \cdot y)$

   i $2 + x \div 3$ j $(2 + x) \div 3$

   k $x - (4 + s) \div w$ l $x - 4 + s \div w$

   m $x \cdot x + 1$ n $(x + 1) \cdot (x + 1)$

   o $a \cdot a - b \cdot b$ p $(a + b) \cdot (a - b)$

   q $(q + b) \div (q - b)$ r $(q + b) \div q - b$

   s $1 \div (2 \cdot s)$ t $1 \div 2 \cdot s = s \div 2$

   u $s \div 2$ v $3 \cdot (x + 1) \div y$

   w $3 \cdot x \div (2 \cdot y)$

   x $3 \cdot x \div 2 \cdot y = 3 \cdot x \cdot y \div 2$

   y $3 \div 2 \cdot x \cdot y = 3 \cdot x \cdot y \div 2$

   z $3 \div (2 \cdot x \cdot y)$

5 a First 2 times s, then plus t

   b First s plus t, then times 2

   c First x minus y, then times 3

   d First 3 times x, then minus y

   e 3 times x times y

   f 3 times x divided by y

   g 2 times x minus 4 times y

   h First x minus 4 times y, then times 2

   i First x divided by 3, then plus 2

   j First 2 plus x, then divided by 3

   k First 4 plus s, then divide by w, then subtract from x

   l First s divided by w, then add to the result of x minus 4

   m First square x, then add 1

   n First x plus 1, then square

   o First square b, then subtract from a squared

   p First sum a and b, then subtract b from a, then multiply the two results

   q First sum q and b, then divide by the difference between q and b

   r First sum q and b, then divide by q, then subtract b

   s 1 divided by the result of 2 times s

   t One half of s

   u s divided by 2

v First x plus 1, then times 3 and divided by y

w 3 times x divided by the result of 2 times y

x First 3 times x divided by 2, then times y

y First 3 divided by 2, then times x and times y

z 3 divided by the result of 2 times x times y

Practice questions 7.2.1

1 a $x + x + 1\ (= 2x + 1)$

b $x + 3$

c $2 + x + 2 = x + 4$

d $3 + x + x + x\ (= 3x + 3)$

e $x + x + 7 = (2x + 7)$

2 a $(x + 1) + (x + x + 2)$

b $(x + x + 2) + (x + x)$

c $(x + 2) + (5)$

d $(x + x + 2) + (0)$

e $(7) + (x + x)$

3 a

b

c

d

e

Practice questions 7.2.2

1 a 2   b 3   c 3

   d 2   e 4   f 3

2 a $2x, 4x$ b $1, 4$

c $w, -4w$ d $2x^2, 4y, -3x^2, y$

e $1, 4$ f $xy, -3xy$

g $2mn, -mn$ h $9n^2, -5mn, 3n^2, 2mn$

i xy, yx

j $2st^2, 3t^2, -4, -t^2, st^2, 1$

k $xyz, yz^2, -2xyz, -2yz^2$

3 a $5x$ b $4y$ c $3a$

d $7t$ e $2w$ f $2x$

g 0 h $3w$ i $3xy$

j x k $3ab$ l $x + 7$

4 a $5x + 3y$ b $-4w + 5t$

c $4x - 7$ d $x + y$

e $3x^2 + 3y$ f $3x^2 + 3x$

g $2x + 3$ h $-2 + 6t$

i $7a^2 + 1$ j $x^2 + 4t$

k $5j + i + i^2$ l $-2b$

m $3a^2 - 2a$ n $ab + a + b$

o $3xy + x^2$

5 a $8x$ b $2n$

Practice questions 7.2.3

1 a 2   b -1 c 5

d 9   e $-\dfrac{1}{3}$ f -11

g 1   h 5   i 5

j 5   k $-\dfrac{4}{3}$

Practice questions 7.2.4

1 a 2   b 3   c 3   d 2

e 3   f 2

2 a $3b + 9$ b $2w + 2t$

c $7x - 7$ d $6 - 2x$

e $-2x - 2$ f $-2c + 6$

3 a $6b + 3c + 9$ b $2w + 4t - 6$
   c $12s + 20t + 4$ d $-4x + 2y - 2$
   e $8 + 12x$ f $6w - 2$
4 a $b^2 + 3b$ b $2wt + 2t^2$
   c $4e^2 - 8e$ d $-x^2 - x$
   e $3x - 9x^2$ f $-6y^2 - 2y$
5 a $6b + 5$ b $2 + 3t$
   c $-1 + 8m$ d $3x - 2$
   e $-4 - 2x$ f 8

Practice questions 7.3

1 a x^3 b w^5
   c x^3y d x^2y^2
   e e^3t^3 f $(ab)^3 = a^3b^3$
   g x^4y h a^4b^2
2 a $r \cdot r$
   b $x \cdot x \cdot x$
   c $a \cdot b \cdot b$
   d $ab \cdot ab = a \cdot b \cdot a \cdot b$
   e $x \cdot y \cdot z \cdot z \cdot z$
   f $xyz \cdot xyz \cdot xyz = x \cdot y \cdot z \cdot x \cdot y \cdot z \cdot x \cdot y \cdot z$
   g $a \cdot a \cdot b \cdot b \cdot b$
   h $x \cdot yz \cdot yz = x \cdot y \cdot z \cdot y \cdot z$
3 a r^5 b x^4 c ab^5
   d a^3b^2 e m^5n^3 f $6x^3y$
   g $18a^3b^2$ h n^3m^5
4 a r b x^2 c y d x^8
   e x^9 f t^8 g x^3 h x^2

Practice questions 7.4

1 a What number added to 1 gives 3? 2
   b From what number do I subtract 3 to obtain 6? 9
   c What number added to 3 gives 4? 1
   d From what number do I subtract 3 to obtain −2?
   e What number do I subtract from 6 to

obtain 2? 4
   f What number added to 4 gives 6? 2
   g What number multiplied by 3 gives 12? 4
   h What number multiplied by 4 gives −16? −4
   i What number multiplied by 7 gives 21? 3
   j What number multiplied by −2 gives 8? −4
2 a $x = 3$ b $p = 1$ c $x = 7$
   d $m = 8$ e $w = 9$ f $x = 0$
   g $d = 3$ h $m = 7$ i $x = 8$
   j $p = 6$ k $y = 11$ l $x = 21$
   m $x = 5$ n $x = 2$ o $x = 6$
   p $x = 5$ q $x = 7$ r $x = 14$
3 a $x = -9$ b $p = -1$ c $x = -7$
   d $m = 22$ e $w = -15$ f $x = 16$
   g $d = -33$ h $m = -7$ i $x = 4$
   j $p = 2$ k $y = -11$ l $x = -3$
   m $x = 21$ n $x = -6$ o $x = 14$
   p $x = -37$ q $x = -7$ r $x = -14$
4 a $x = 4$ b $x = \dfrac{9}{2}$ c $x = 4$
   d $x = 9$ e $x = -1$ f $x = 4$
   g $x = 0$ h $x = 6$ i $x = 4$
5 a $x = 3.1$ b $x = -0.2$ c $x = 3.3$
   d $x = 3$ e $x = 7.5$ f $x = 10$
   g $x = -0.6$ h $x = 1.25$ i $x = -1$
   j $x = 2$ k $x = 3$ l $x = 13$
6 a $x = 1$ b $x = -1$ c $x = -1.5$
   d $x = 1$ e $x = \dfrac{3}{4}$ f $x = \dfrac{7}{2}$
   g $x = -8$ h $x = 0$

Practice questions 7.5.1

1 a $3m$ b $P + 4$ c $n + (n + 1)$
   d $2A$ e $F + 2$ f $B - 1$
   g $b - h$ h $\dfrac{bh}{2}$ i $\dfrac{D + d}{2}$
2 a $x + 4 = 10 \Rightarrow x = 6$
   b $x - 5 = 7 \Rightarrow x = 12$
   c $x - 4 = 1 \Rightarrow x = 5$

d $3x = 12 \Rightarrow x = 4$

e $\dfrac{x}{4} = 6 \Rightarrow x = 24$

f $\dfrac{20}{x} = 5 \Rightarrow x = 4$

g $x + 5 = -3 \Rightarrow x = -8$

h $5 - x = 0 \Rightarrow x = 5$

i $4 - x = -1 \Rightarrow x = 5$

j $5x = 2 \Rightarrow x = \dfrac{2}{5}$ or 0.4

k $\dfrac{x}{3} = -4 \Rightarrow x = -12$

l $\dfrac{-10}{x} = 5 \Rightarrow x = -2$

3 a $3x - 8 = 1 \Rightarrow x = 3$

b $4 + 2x = 8 \Rightarrow x = 2$

c $\dfrac{x}{3} + 8 = 10 \Rightarrow x = 6$

d $50 - 4x = 30 \Rightarrow x = 5$

4 a $4n = 32 \Rightarrow n = 8$

b $5c = 25 \Rightarrow c = 5(\$)$

c $3a + 2p = 12$ and $p = 3 \Rightarrow a = 2(€)$

d $n + (n + 1) = 11 \Rightarrow n = 5, n + 1 = 6$

e $n + (n + 2) = 22 \Rightarrow n = 10, n + 2 = 12$

f $n + n + 2 + n + 4 = 39 \Rightarrow 3n \Rightarrow n = 11$ and 13 and 15 (odd numbers)

g $3s + 2d = 12, s = 2d \Rightarrow d = 1.5(\$)$

h $3s + 2d = 12, s = d + 0.5 \Rightarrow d = 2.1(\$)$

5 30

Practice questions 7.5.2

1 a

x	0	1	2	3	4
LHS, $3x + 1$	1	4	7	10	13
RHS, $9 - x$	9	8	7	6	5

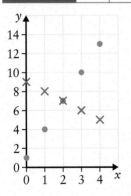

$x = 2$

b

x	−1	0	1	2	3	4
LHS, $3 - 2x$	5	3	1	−1	−3	−5
RHS, $x - 6$	−7	−6	−5	−4	−3	−2

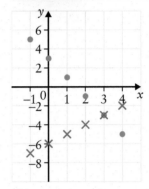

$x = 3$

c

x	−1	0	1	2	3	4
LHS, $x + 5$	4	5	6	7	8	9
RHS, $10 - x$	11	10	9	8	7	6

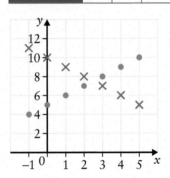

$2 < x < 3$

d

x	−1	0	1	2	3	4
LHS, $3(x + 1)$	0	3	6	9	12	15
RHS, $-2(x - 9)$	20	18	16	14	12	10

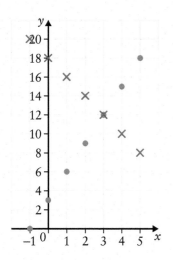

$x = 3$

e

x	−1	0	1	2	3	4
LHS, $2x − 1 + 3(x − 1)$	−9	−4	1	6	11	16
RH, $2x + 2$	0	2	4	6	8	10

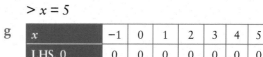

$x = 2$

f

x	−1	0	1	2	3	4	5
LHS, $x + 5$	4	5	6	7	8	9	10
RHS, 10	10	10	10	10	10	10	10

> $x = 5$

g

x	−1	0	1	2	3	4	5
LHS, 0	0	0	0	0	0	0	0
RHS, $−3 + x$	−4	−3	−2	−1	0	1	2

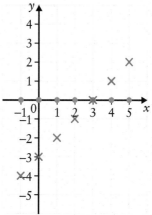

$x = 3$

h

x	−2	−1	0	1	2	3
LHS, $2(x + 3)$	2	4	6	8	10	12
RHS, $3x + 6$	0	3	6	9	12	15

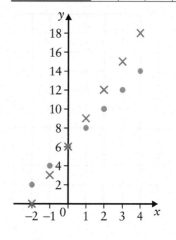

$x = 0$

i

x	−2	−1	0	1	2	3	4
LHS, $3 + 2(x − 4)$	−9	−7	−5	−3	−1	3	5
RHS, $x − 5$	−7	−6	−5	−4	−3	−2	−1

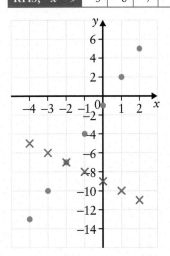

$x = 0$

j

x	−4	−3	−2	−1	0	1	2
LHS, $3x − 1$	−13	−10	−7	−4	−1	2	5
RHS, $−x − 9$	−5	−6	−7	−8	−9	−10	−11

$x = −2$

k

x	−4	−3	−2	−1	0	1
LHS, $x + 1$	−3	−2	−1	0	1	2
RHS, $−5 − x$	−1	−2	−3	−4	−5	−6

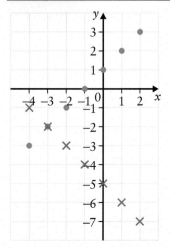

$x = −3$

l

x	−4	−3	−2	−1	0	1
LHS, $3(x − 4) + 1$	−10	−8	−6	−4	−2	0
RHS, $2(x + 1) + x$	−6	−4	−2	0	2	4

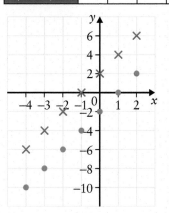

no solutions

m

x	−3	−2	−1	0	1	2
LHS, $3(x-4)+14$	−7	−4	−1	2	5	8
RHS, $2(x+1)+x$	−7	−4	−1	2	5	8

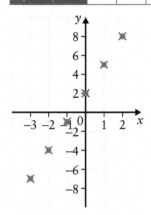

x = any number

2 a i

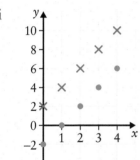

ii

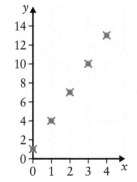

b Part i has no solutions because the two lines are parallel.

Part ii has infinitely many solutions because the two lines overlap.

Check your knowledge questions

1 a $7x + 17$ b $-6x - 14$
c $5x + 5$ d $-9x + 15$
e $-5x + 11$ f 0
g $4a + 3b$ h $3a - 2b$
i $-2xy - 4x$ j $2x$
k $-3x^2 + 3x$ l $2axy$

2 a 5 b -5 c 1
d 5 e 1 f 2
g 1 h -2 i -4
j 4 k -0.5 l -1

3 a $x = 5$ b $r = 8$ c $w = 2$
d $x = 17$ e $x = 3$ f $x = 2$
g $x = 0$ h $x = -2$ i $x = -3$
j $x = -7$

4 a $x = 4$ b $x = 2$ c $x = 3$
d $x = 16$ e $x = 12$ f $x = 30$
g $x = -3$ h $x = -5$ i $x = -10$
j $x = -8$ k $x = -21$ l $x = 8$
m $x = 3$ n $x = -4$ o $x = -2$

5 a $x = 4$ b $x = 6$ c $x = -4$
d $x = -1$ e $x = 4$ f $x = -1$
g $x = 2$ h $x = 2$

6 $x = 8$ cm

7 21 cm

8 Base = 20 cm, height = 10 cm

9 27, 28

10 $100

11 a $17   b 10 hours

12 a £167   b 206 times

13 $6.00

14 a Beryllium   b Lithium
c Helium and helium

15 a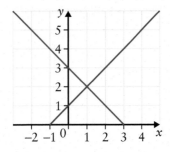

$x = 1$

b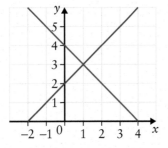

$x = 1$

c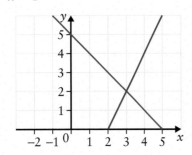

$x = 3$

d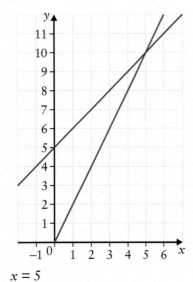

$x = 5$

e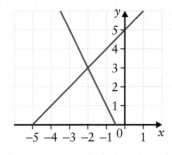

$x = -2$

f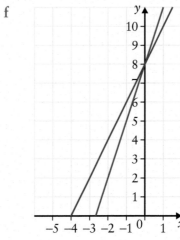

$x = 0$

Chapter 8 answers

Do you recall?

1 a Rectangle

   b Square

   c Parallelogram

   d Rhombus

   e Trapezium

   f Isosceles trapezium

   g Kite

2 a Cuboid   b Cube

3 a $4x$ b $3y$ c $2x + 5y$

4 a 7   b 365   c 366

5 24

6 60

7 60

8 a 8 a.m.   b 5 p.m.

c 7.30 p.m.　　　　d 12.28 p.m.

e 12.15 a.m.

9 a 11:15　　　　　b 01:40

　c 15:10　　　　　d 22:30

10 a 6.45 p.m.　　　b 18:45

Practice questions 8.1.1

1 A a 8 mm　　　　b 0.8 cm

　B a 22 mm　　　　b 2.2 cm

　C a 45 mm　　　　b 4.5 cm

　D a 80 mm　　　　b 8 cm

　E a 118 mm　　　b 11.8 cm

2 A a 12 cm　　　　b 120 mm

　B a 14.5 cm　　　b 145 mm

　C a 16.7 cm　　　b 167 mm

　D a 17.9 cm　　　b 179 mm

　E a 19.3 cm　　　b 193 mm

3 a 70 mm　　b 48 mm　　c 1420 mm

　d 300 cm　　e 1400 cm　　f 250 cm

　g 2000 mm　h 1600 mm　i 26 000 m

　j 5700 m

4 a 4.9 cm　　b 12 cm　　c 4 m

　d 0.96 m　　e 3 m　　　f 5 km

　g 0.375 km　h 0.3 km

5 a 3600 cm　b 4.9 m　　c 70 cm

　d 2300 mm　e 12 000 m　f 1200 mm

　g 2.7 km　　h 47.5 m

6 0.01 km, 574 cm, 5470 mm, 5.2 m, 0.005 km

7 1.7 m

8 40 lengths

9 6 shelves

10 Petra, by 100 m

11 a No, width is 545 mm

　b No, height is 940 mm

12 1500 cm

Practice questions 8.1.2

1 7 m 51 cm 7 mm

2 a 324.5 cm　　　　b 3.245 m

3 A 2.100 m　　　　B 2.166 mm

4 a i Metres　　　ii 5 m tape measure

　b i Centimetres　ii Ruler or tape measure

　c i Millimetres　ii Ruler

　d i Centimetres

　　ii Metre rule or tape measure

　e i Metres　　　ii 5 m tape measure

　f i Metres　　　ii Trundle wheel

　g i Kilometres

　　ii Odometer or trundle wheel depending on distance

5 a 5 cm　　b 7 cm　　c 6 cm

　d 6 cm　　e 3 cm

6 a 4.8 cm　b 7.2 cm　c 5.5 cm

　d 6.5 cm　e 3.3 cm

7 2.7 m by 3.1 m

8 106.2 m and 68.8 m

9 66.7 cm

10 18.9 km

11 8.5 cm, 8.7 cm, 8.9 cm, 9.1 cm, 9.4 cm

12 The wall is between 2.25 m and 2.34 m long to the nearest cm, so the 228 cm sofa may be longer than the wall.

13 a 53 096.4 km

　b The car may not have travelled exactly 53 096.4 km. It could have travelled, e.g. 53 096.448 km. Another 3600 m would then give a total of 53 100.048 km

Practice questions 8.1.3

1 a 18.4 cm　　　　b 12.6 cm

　c 11.1 cm　　　　d 10.8 cm

2 a 10.6 cm　　　　b 11 cm

3 a 20 cm   b 20 cm

4 a 15.9 cm   b 24 cm

   c 54 cm   d 120 mm = 12 cm

5 12.3 cm (to the nearest millimetre)

6 4.8 cm

7 a 50 m   b 40 m

8 22 cm

9 163 mm = 16.3 cm

10 11.5 cm

11 31 cm

12 40 rolls = $639.60

13 56 cm

Practice questions 8.2.1

1 a 10 cm^2 b 19 cm^2 c 15 cm^2

2 a $\frac{1}{2}$ cm^2 b $\frac{1}{2}$ cm^2

   c $\frac{1}{4}$ cm^2 d $\frac{3}{4}$ cm^2

3 a 6 cm^2 b 4 cm^2 c 6 cm^2

4 14 cm^2

5 area = 6 cm^2 perimeter = 14 cm

6 a 50 cm^2 b 120 mm^2 c 153 m^2

7 49 cm^2

8 11 cm

9 3.2 cm

10 a 180 mm^2 b 1.5 m^2

    c 10 500 cm^2

11 a 30 cm^2 b 40 cm^2 c 28 cm^2

12 4 cm

13 a 72 cm^2 b 70 cm^2

    c 87 mm^2 d 73 cm^2

14 a 66 cm^2 b 58 mm^2

15 $117.00, 9 square metres required

16 77 m^2

17 6 m by 6 m square

18 1 m by 18 m

Practice questions 8.2.2

1 a 25 cm^2 b 42 cm^2 c 90 cm^2

   d 28 cm^2 e 49.5 cm^2 f 21 cm^2

2 a 16 cm^2 b 40 cm^2 c 17.5 cm^2

3 Area 30 cm^2, perimeter 30 cm

4 a 65 cm^2 b 32 cm^2 c 38 cm^2

5 320 mm^2

6 4 cm

7 136 cm^2

8 38 cm^2

9 108 m^2

10 24 litres, so 5 cans needed. $80

11 50 cm^2

12 Student's own accurate drawings, e.g.

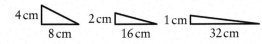

Practice questions 8.3

1 a 8 cm^3 b 15 cm^3 c 14 cm^3

2 a 42 cm^3 b 30 cm^3

3 a 210 cm^3 b 2520 m^3 c 1980 mm^3

4 27 cm^3

5 125 cm^3

6 a   b 60 cm^2

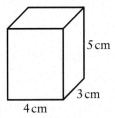

7 3840 cm^3

8 45 m^3

9 4 cm

10 7 cm

11 Volume is 109 440 cm^3, which is greater than 100 000 cm^3

12 a 180 cm^3 b 96 cm^3

13 420 cm³

14 a Student's own sketches, e.g. cuboids with dimensions 1, 1, 24 cm; 2, 4, 3 cm

    b Student's own sketches, e.g. cuboid with dimensions 1.5, 4, 4 cm

15 Student's own answers

16 5 cm

Practice questions 8.4.1

1 a 12 March   b 2 March

   c 23 March

2 21 November

3 6 November

4 a 30   b 15   c 45   d 30

   e 80   f 3 hours 45 minutes

   g 1 minute 38 seconds

5 a 1440   b 3600   c 168

6 $\dfrac{40}{168} = \dfrac{5}{21}$

7 24 days

8 3.15 p.m.

9 12.06 p.m.

10 a Funny film   b 35 minutes

11 12.05 p.m.

12 11.35 a.m.

13 11 hours 20 minutes

14 7.40 a.m.

15 Student's own answers

Practice questions 8.4.2

1 a 25 minutes   b 35 minutes

   c Documentary

2 a 54 minutes

   b 15 minutes from the park to the rail station

   c 28 minutes

   d 27 minutes

   e 1550, 1628, 1714

3 a 13 hours 35 minutes

   b 1 hour 21 minutes

   c 1449 and 1849

   d 2 hours 21 minutes

4 7 hours 46 minutes

5 a 0808   b 8 minutes

   c 30 minutes   d 9.58 a.m.

   e 15 minutes

6 1750 or 5.50 p.m.

7 1530 or 3.30 p.m.

8 5 hours 58 minutes

Check your knowledge questions

1 a 42 000 cm   b 5.3 m

   c 400 000 mm   d 54.6 cm

   e 2.45 km   f 3400 mm

2 55 books

3 a mm, ruler

   b m, measuring tape

   c km, trundle wheel

   d m, measuring tape/trundle wheel

   e Seconds, stopwatch

4 28.6 cm

5 Between 7.2 m and 8 m

6 a 18.3 cm   b 106 mm

   c 30.8 m   d 40 cm

7 63 cm

8 2.5 cm

9 11 cm²

10 40 cm²

11 a 108 mm²   b 14 cm²

    c 57 m²   d 14 cm²

12 a 8.5 cm   b 8 cm

13 162 cm³

14 40 cm³

15 1725

16 470 minutes

17 a Every 35 minutes

   b 1306   c 7 minutes

18 12 noon (1200)

Chapter 9 answers

Do you recall?

1 a 6   b Blue

2 a 4.8   b 10.6

3 a 12   b 4.5

4 73°

5 a 23   b 15

6 a $\frac{13}{20}$ b 65%

Practice questions 9.1

1 a Elli's sample includes only students at her school. This sample will not be representative of the whole population, which, in this case, is the community.

   b Elli could ask a randomly selected group of people in her community. She should aim for the largest sample size possible, so that the results are meaningful.

   c Not all music choices are represented in Elli's questionnaire.

   d Elli should add other categories, also include 'Other', and include space for people to write in what their favourite type of music is, if it is not included on the list.

2 a Michelle carried out a census. She collected data from every individual in the population.

   b Matteo's sample will almost certainly be the truest representation of the students at the school, because his sample included students from each grade.

c David had asked only students from the playground. Students who were not on the playground when he conducted the survey would not be included. Also, his sample size was too small to produce meaningful results. Sophie's sample size was good, but she surveyed only one class, so students in all four of the other grades would not be represented.

3 a No, Hakan has not rolled the dice enough to make such a conclusion.

   b Hakan should roll the dice many more times in order to decide whether or not it is biased. Suggest: 60 rolls

   c This is similar to having a sample size of just 8, which will never be enough from which to draw meaningful conclusions.

4 The poll does not include people who do watch TV but who do not read the magazine.

5 13 from MYP 1, 16 from MYP 2, 19 from MYP 3, 16 from MYP 4, and 16 from MYP 5

Practice questions 9.2

1 a Frequencies (top to bottom): 7, 11, 5, 3, 2, 2

   b 30

   c 1

   d 4

2 a Ordinal – this is like a ranking

   b Discrete – only specific values are allowed

   c Categorical – describes a type of fossil fuel

   d Discrete – when counting grains of sand only whole numbers are allowed

   e Continuous – a percentage is a real number, e.g. a decimal is possible

   f Continuous – a length can be measured to be any real number

g Continuous – it depends how the household waste is measured; discrete if in bins or bags or continuous if by weight

3 a

Shape	Frequency
square	6
oval	4
circle	7
star	10
triangle	7

b 34   c Star

d $\frac{7}{34} \approx 20.6\%$ (1 d.p.)

4 a 10   b 4

c

Correct answers	Frequency
4	1
5	3
6	4
7	7
8	6
9	6
10	3

d 7   e 8

f 9 students, which is 30%

Practice questions 9.3.1

1 a US dollars ($)   b $29

c $37   d 25

2 a 4   b 3   c 9   d 15

e 39

f 21 (7 households with 3 people each)

g 162

3 a

Marks	Frequency
0	3
10	6
20	7
30	10
40	5
50	8
60	0

b 39   c 16

4 a Highest = 33, lowest = 27

b

Candies	Frequency
27	2
28	4
29	6
30	20
31	10
32	6
33	2

c 12   d 18

e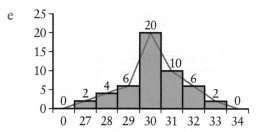

5 a

Response	Frequency
A	28
B	46
C	21
D	5

b 28   c 5%   d 74%   e 26%

f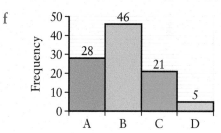

6 a 2007

b 4.5 Mha

c 64%

Practice questions 9.3.2

1 a 30 pineapples   b 5 pineapples

c 15 pineapples   d 55 pineapples

2 a 11   b 5

c

3 a Student's own graph

b Student's own reason(s), e.g. it is better to use a bar chart here because it is easier to see the difference between frequency values of 8, 9 and 10 rather than in areas of sectors in a pie chart

4 170.55 megatonnes

5 a $\frac{1}{3}$ b 5 cars

6 a 32 b 64

7 a Manufacturers and processors: 154°

Transport: 7°

Cocoa farmers: 23°

Taxes: 15°

Retailers: 159°

b Retailers

c 6.6%

8 a 10°

b

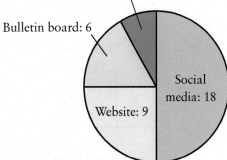

9 a

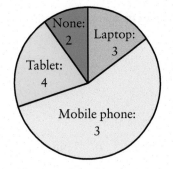

b 55%

10 a

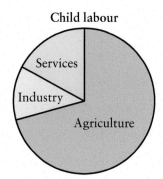

b Africa 25.4%

Americas 6.7%

Asia-Pacific 9.2%

Europe and Central Asia 5.1%

All other countries 53.6%

11 Agriculture 71%   256°

Industry 12%   43°

Services 17%   61°

Child labour

12 Democratic Republic of Congo 61%   219.6°

   Asia-Pacific 16%   57.6°

   Africa 8%   28.8°

   Europe 6%   21.6°

   Latin America 5%   18°

   USA & Canada 4%   14.4°

13 a i False   **ii** True

   iii Cannot tell from the chart

b The number of businesses represented in each chart

Practice questions 9.4

1 a

Stem	Leaf
1	5 6 7 8 9
2	6 6 6 8 9
3	0 3 5 6 8
4	0 1 7 7 7
5	1 8
6	2 2

Key: 1|5 = 15

b

Stem	Leaf
4	2 4 5 9
5	3 7 9
6	0 3 5 7
7	0 2 3 4 7
8	0 0 1 1 2 8
9	1 2

Key: 4|2 = 42

2 a

Stem	Leaf
0	1 2 4 6 8 9
1	1 5 8 9
2	0 6 7 9
3	2 3 6
4	3 5 6 9
5	
6	2

Key: 2|0 = 20

b Student's own tables

c 45% (nearest percentage)

3 a

Lithium ion	Stem	Lithium polymer
8	1	
2	2	
6 6 2 2	3	6 8 8
8 6 4 2 0	4	0 0 2 4 6 8
4 2 2 0	5	0 2 5 6 6
4 4 2 2 0	6	0 0 0 0 6 8
6 4 0	7	0 0 4 8
4 4 4 2 0	8	0 2 2 4 4 8
	9	0 4

Key: Lithium ion: 2|3 = 32
   Lithium polymer: 5|0 = 50

b Student's own interpretations, e.g. lithium polymer has higher values as in greater than 80 hours; lithium ion batteries has values less than 20 hours.

4 a

Children	Stem	Adults
	1	2
	2	
	3	
	4	
	5	5 6 7 8 8 9
7 2	6	0 0 1 4 4 4 8 8
8 7 7 2	7	2 2 2 6 6
8 6 6 6 5 5 2 1	8	0 0 1
9 9 8 5 4 4 2	9	2
2 0	10	

Key: Children: 2|6 = 62
   Adults: 8|0 = 80

b Student's own interpretations, e.g. children tend to have higher values than adults

Practice questions 9.5

1 a No relationship

 b Positive relationship

 c Negative relationship

 d Negative relationship

 e Positive relationship

2 a i Akeel: Maths = 5, English = 7

 ii Luke: Maths = 20, English = 19

 iii Amy; Maths = 9, English = 13

 iv Ramisha: Maths = 18, English = 14

 b i Highest maths mark = 20

 ii Highest English mark = 19

 c Positive relationship

3 a

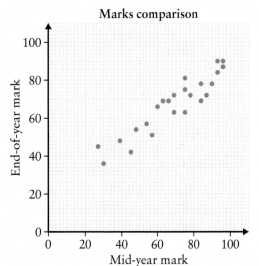

Marks comparison

 b Positive relationship. The Mid-year results increase as the End-of-year results increase.

Check your knowledge questions

1 a

Number of gym visits	Tally	Frequency
0	IIII	4
1	IHtI	6
2	II	2
3	II	2
4	I	1
5	II	2
6	II	2
7	I	1

 b

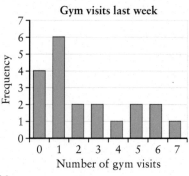

Gym visits last week

 c 20

 d Student's own conclusions, which could include:

 50% of the people asked went to the gym less than twice in the week

 One very keen person reported going every day!

2 a There is no option for 'did not exercise at all last week'; the 3–4 times and 4–5 times options overlap.

 b e.g. How many times did you exercise last week?

 none ☐   1–2 times ☐   3–4 times ☐

 5–6 times ☐   more than 6 times ☐

3 a 100   b 30   c $\frac{30}{100} = \frac{3}{10}$

4 a January, February and March

 b i February   ii June

 c April

5

Music	30°
Food and drink	100°
Clothes	50°
Phone contract	120°
Other	60°

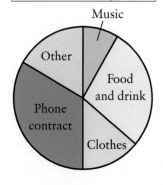

6 Student's own dual bar charts

7 a Winter b 17 000 c 22 500

   d Summer: 16 000 + 17 000 = 33 000 visitors

      Winter: 4000 + 6500   = 10 500

      Spring: 9000 + 11 000 = 20 000

                              30 500

      Thus, there were more visitors in the
      summer (33 000) than in winter and spring
      combined (30 500).

8 Student's own pictogram

9 Student's own answers

10 a

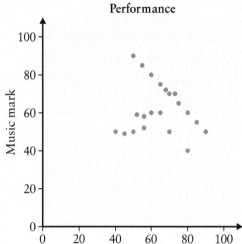

   b Student's own comments, e.g. there
      appears to be a negative relationship